Logic Pro X

Logic Pro X
Audio and Music Production

Mark Cousins

Russ Hepworth-Sawyer

Focal Press
Taylor & Francis Group

NEW YORK AND LONDON

First published 2014
by Focal Press
70 Blanchard Road, Suite 402, Burlington, MA 01803

and by Focal Press
2 Park Square, Milton Park, Abingdon, Oxon OX14 4RN

Focal Press is an imprint of the Taylor & Francis Group, an informa business

Library of Congress Cataloging in Publication Data
CIP data has been applied for.

ISBN: 978-0-415-85768-0 (pbk)
ISBN: 978-0-203-79660-3 (ebk)

Typeset in Myriad Pro
Project Managed and Typeset by: diacriTech
Printed in the USA

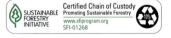

Bound to Create

You are a creator.

Whatever your form of expression — photography, filmmaking, animation, games, audio, media communication, web design, or theatre — you simply want to create without limitation. Bound by nothing except your own creativity and determination.

Focal Press can help.

For over 75 years Focal has published books that support your creative goals. Our founder, Andor Kraszna-Krausz, established Focal in 1938 so you could have access to leading-edge expert knowledge, techniques, and tools that allow you to create without constraint. We strive to create exceptional, engaging, and practical content that helps you master your passion.

Focal Press and you.

Bound to create.

We'd love to hear how we've helped
you create. Share your experience:
www.focalpress.com/boundtocreate

Focal Press
Taylor & Francis Group

Dedication

This edition of our book is dedicated to the memory of Richard "Rich T" Tolson (1973–2013), a dedicated music educator, friend, and colleague to many. The music world is poorer without him.

Table of Contents

About the Authors

Mark Cousins works as a composer, programmer, and sound engineer (www.cousins-saunders.co.uk), as well as being a long-serving contributor to *Music Tech* magazine. His professional work involves composing music for some of the world's largest production music companies – including Universal Publishing Production Music, among others – and he has had his music placed on major campaigns for brands such as Strongbow, McDonalds, Stella Artois, Hershey, BT, and Liptons as well as being used on hit shows such as The Apprentice, Top Gear, and CSI:NY. He has also had works performed for the Royal Philharmonic Orchestra, the East of England Orchestra, City of Prague Philharmonic Orchestra, and the Brighton Festival Chorus.

Mark has been an active contributor to *Music Tech* magazine since issue one. He has been responsible for the majority of cover features, as well as the magazine's regular Logic Pro coverage. As a senior writer, he has also had a strong editorial input on the development of the magazine, helping it become one of the leading brands in its field.

Russ Hepworth-Sawyer is a sound engineer and producer with two decades' experience of all things in audio and is a member of the Association of Professional Recording Services; a Fellow of the RSA (Royal Society for the encouragement of Arts, Manufactures and Commerce), and a board member of the Music Producer's Guild and helped form their Mastering Group. Through MOTTOsound (www.mottosound.co.uk), Russ works freelance in the industry as a mastering engineer, writer, and consultant. Russ currently lectures part time for York St John University and has taught extensively in Higher Education at institutions including Leeds College of Music, London College of Music, and Rose Bruford College and has contributed sessions at Barnsley College. He has written for *MusicTech* magazine, *Pro Sound News Europe*, and *Sound on Sound* plus has written many titles for Focal Press.

Acknowledgments

I would like to thank my mum and dad for gifting me a Casio CZ-3000 as my Christmas present in 1987 and starting the whole ball rolling; my wife, Hannah, and our two kids, Josie and Fred, for their support, patience, and amusement over the years and for keeping me company in the studio from time to time; Russ, who has been a longtime and sincere friend, but whose suggestion of writing a book together I should have politely declined! Neil Worley, for taking me on at *Music Tech* and showing me the hallowed profession of a being a writer; Adam Saunders, for putting up with my lack of musical output for several months; and finally, Polly, Deb, and Cath for looking after Fred in those final few hectic weeks!

Mark Cousins

I would like to thank my wife Jackie and sons Tom & Aaron for their continued support as another book deadline loomed – thank you!; my parents-in-law Ann and John for all their help; Mark for his guidance, knowledge, and for being the top chap he is; Max Wilson for his endless support and forcing me to move to Logic in the first place … thanks boss!; Jon Miller for the use of his animation file in Chapter 10, wherever he is! Iain Hodge, Peter Cook, and Justin Paterson at the London College of Music; and Craig Golding, and Danny Cope at Leeds College of Music. Thanks too to colleagues at York St John, namely Rob Wilsmore, Ben Burrows, Dave Young, Andy Devine, and Andy FarThing. Thanks too to Laura for propping MOTTOsound up whilst I've been concentrating on this. Thanks also go to Barkley McKay with whom I've had many DAW related "discussions" … Barkley McKay (one day you'll use Logic too!), plus hello to the Hepworths, Sawyers, and Leathers out there – you know who you are….

Russ Hepworth-Sawyer

In this chapter

Knowledgebases

The Logic Pro Concept

1.1   Introduction

It's hard to imagine a more complete system for music and audio production than Logic Pro X: multitrack recording and editing, a full suite of virtual instruments and effects, and a seamless workflow that takes you from the beginning of your project right through to the delivery of the final production master. Yet, with such a complete system comes the daunting task of understanding how the elements of Logic Pro X knit together to produce a professional-sounding result. For example, where do you begin to start writing music or making a recording in Logic? What are the virtual instruments and plug-ins used by the professionals to create release-quality results? And how can you transform those poorly performed band recordings into a polished CD?

So, let's be clear from the start: this book isn't just another instruction manual for Logic Pro X. Instead, we've taken a process-driven approach that appraises, understands, and explores the features of Logic Pro X in a way that matches the structure and order of the production process. We'll do more than just technically describe the functions of Logic Pro X; we'll look at how the various elements of Logic Pro X relate to the demands of audio and music production. With all but a few exceptions, most of the chapters focus on a specific part of the production process – whether it's initial track laying, sound design, or mastering your finished mixes to produce the final CD – highlighting the relevant parts of the application that guarantee a professional-sounding audio product. We'll also look at techniques that go beyond the scope of the manual – practices like parallel compression, for example, that many engineers use and abuse on a daily basis.

If you're starting off from scratch, it's easy to be overwhelmed by the sheer size and complexity of an application like Logic Pro X. However, it isn't essential to understand the entirety of the application to start producing music. Get to know the components that are most relevant to your way of working and build from there – use plenty of presets, Apple Loops, and so on to get you kick started – and then enjoy the process of exploring each element that little bit further. Ultimately, Logic Pro X is a tool that will grow with your experience – a system that will surprise at every turn and open up new possibilities whenever you want to explore the software further. With this book, you'll at least have a reference to aid you in that process, but don't be afraid to experiment to find out how Logic Pro X best fits into your unique creative process!

1.2   A Brief History of Logic Pro X

Like the other "old-timers" of computer-based audio production – including Cubase and Pro Tools – Logic Pro X is an application with a rich and long heritage in the music industry. Born from the ashes of C-Lab's Notator and Creator in 1993, Notator Logic (as it was then called) was an attempt to create a visual, region-based production environment for MIDI sequencing. Building blocks, or regions of MIDI data – used to control hardware synthesizers and samplers – could be arranged on the computer screen, with a

Figure 1.1 Logic Pro X includes an impressive array of features, but understanding how they integrate into the production process might not be immediately apparent.

clear, visual representation of the structure of the arrangement. What was unique about Logic, though, was that the application was completely configurable – users could create virtual presentations of their studio, known as an environment, for example, or combine different editor windows in a completely configurable user interface.

Audio functionality was added to the application in 1994, with the release of version 1.7, allowing Logic users to combine both digital audio and MIDI data all in the same arrangement (although initially, only with expensive Digidesign audio hardware). Virtual instruments followed in 2000, making the system a complete production environment where music could be composed, mixed, and mastered all in one computer, and arguably, without the need for any extra third-party software. Although revolutionary at the time, this method of production has now become the norm, with many musicians and engineers largely working entirely "in the box."

Apple acquired the company that originally developed Logic – Emagic – in 2002, with its programming team joining Apple's, and Logic Pro becoming part of Apple's prized suite of media-based applications, including Final Cut Studio and Aperture 2. The partnership led to many of Logic Pro's technologies migrating into other Apple applications – most notably with the introduction of GarageBand – as well as Apple making Logic Pro an increasingly price-competitive option, with both the absorption of previously optional software components into the main application (like Space Designer, the EXS24 Sampler, and the ES2 synthesizer), and, with the release of Logic Studio, a halving of its retail price.

Over the development of Logic Pro 8, Logic Pro 9, and now Logic Pro X, Apple has made some big developments in respect to making the application easier and more intuitive to use, yet at the same time, retaining the advanced features that have made Logic so popular with professional musicians and sound engineers. Logic Pro X is arguably the biggest cosmetic change to date, with a completely refreshed user interface and plenty of powerful new features.

1.3   The Logical Advantage

Opinion and debate will always rage as to the "best" digital audio workstation, but there are a number of factors that give Logic Pro X the edge over alternative solutions. Certainly, if you're trying to make a decision between different audio applications – all with such a compelling range of features – it's well worth understanding some of their main overriding benefits, as well as seeing whether these align with your intended method of working.

Complete Integration with Apple Hardware and Software

Being part of Apple, you can guarantee that Logic Pro X will make optimal use of both Apple's computing hardware and the operating system that ties it all together. For example, where other developers might lag behind certain OS updates, Logic Pro tends to be first off the block supporting major upgrades and technologies such as 64-bit architecture. On top of this, Logic Pro has always stood out from the crowd in terms of its efficient use of DSP resources, suggesting a well-coded audio engine, as well as plenty of integrated components – for example, the EXS24 sampler or Space Designer reverb – that ensure a completely optimal use of your computer's processor.

Exhaustive Range of Plug-Ins and Instruments

Logic Pro X's integral range of instruments and effects is easily the most comprehensive set available in any off-the-shelf Digital Audio Workstation (DAW). In addition to standard studio stalwarts such as compression, reverb, and equalization, Logic Pro X includes a number of contemporary effects and plenty of software instruments covering everything from vintage Hammond organs to cutting-edge component-modeling synthesizers. Ultimately, with such a diverse collection of tools, you can easily produce a professional, release-quality output without having to resort to additional third-party plug-ins and effects, although, of course, there's no reason why you can't add these at a later point should you wish to do so. As the old saying goes, the only limit with Logic Pro X is your imagination.

KNOWLEDGEBASE 1
Purchasing and Installing Logic Pro X

As with a range of applications nowadays, Logic Pro X is only available through Apple's App Store. Once you've paid for the App, there's an initial download of around 650 MB before you can start using Logic Pro X. When the installation is complete, Logic Pro should appear on the Launchpad or you can also find it via Applications folder on your hard drive.

Technically speaking, the initial download of Logic only covers a "bare bones installation," with a large amount of additional content, particularly in respect to additional sounds and presets, found in a supplementary download. To activate the additional download, go to the menu item Logic Pro X > Download Additional Content. The following screen should then indicate both the contents you've installed and any remaining items that haven't been downloaded.

If you're trying to keep your drive usage to a minimum, there's a lot of sense in being prudent with the amount of content that you're downloading. The biggest use of space is found under the Drum Kit content package, although this is the data used by Drummer, which is a worthy and useful new feature of Logic Pro X. If you're not interested in content compatibility with older version of Logic, the Legacy and Compatibility is probably worth not installing as this can save over 10 GB.

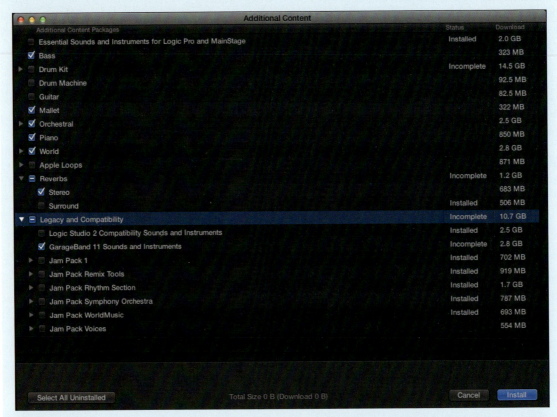

Figure 1.2 Once you've downloaded the initial install of Logic Pro X from the App store, go to the menu option Logic Pro X > Download Additional Content to finish the remaining sound content.

KNOWLEDGEBASE 2
Activating Advanced Tools

To make Logic Pro X easier for newcomers, Apple has thoughtfully given the application two different modes of operation. With the advanced tools active, Logic Pro X users have full access to the functionality of the program. With the advanced tools deactivated, though, Logic Pro X operates with a slimmed-down functionality that's much closer to GarageBand. If you're upgrading from an existing version of Logic Pro 8 or 9, or a similar version of Logic Express, the advanced tools are turned on by default. New users to Logic Pro X, though, will have the advanced tools turned off by default.

Given that the book doesn't make a distinction between advanced tools and a basic toolset, it makes sense to ensure that the advanced tools are turned on. To see the current settings, go to the menu item Logic Pro X > Preferences > Advanced Tools. Check the Show Advanced Tools option to have the full feature set, alongside a series of other additional options that can be dropped in or out as you see fit. While the advanced tools might make the application slightly more daunting, their usefulness in the long run is worth the extra investment of time and effort!

Figure 1.3 Make sure you've activated the advanced tool set (Logic Pro X > Preferences > Advanced Tools) as this will give you full and complete access to Logic's functionality.

Effective Combination of MIDI and Audio Editing

While some applications have strengths in a particular area (like MIDI or audio), Logic Pro X presents an effective hybrid solution for both MIDI-based composition and audio recording and editing, which probably explains why so many working composers operate completely within the realms of Logic Pro X. Its flexibility also makes it possible to use Logic Pro X in a wide range of audio production environments including 5.1 and surround sound mixing, sound for film, and other multimedia applications.

In this chapter

Knowledgebase

Logic Pro X's Interface

2.1 Introduction

Negotiating the interface of any audio application is vital to understand how it works and the precise method involved in creating a finished audio recording. Logic Pro's interface is no exception, providing its own unique slant on the production process: with a number of editor windows, mixers, media browsers, and so on, all used to sculpt and refine your audio output. Navigate Logic Pro's interface in an informed way, therefore, and the application will become a seamless and enjoyable part of your audio production workflow, rather than a stumbling block to your creative exploits.

In this chapter, we're going to take a look at both the overarching principles of using Logic Pro, alongside the specific components of the Logic Pro interface and how these integrate into the production process. From this important stepping-stone, we can then begin the process of creating our own projects and taking a more detailed look at Logic Pro's role and input in audio and music production.

2.2 What Logic Pro Can Record

Logic Pro principally works with two types of data – audio and MIDI. Physical instruments and external sound sources are recorded directly into Logic Pro as audio sound files. You might, for example, plug your guitar straight into a DI input on your audio interface or set up a number of microphones connected to your interface's mic preamps – all of which will be recorded as audio files directly into Logic Pro's main window. Once recorded, you can edit and mix these files to create your finished track, ready to be burned on to CD.

As an alternative to audio files, you can also record MIDI data directly into Logic Pro using an attached MIDI keyboard, which, in turn, can be used to control Logic's integral virtual instruments (or other third-party Audio Unit instruments), as well as external hardware MIDI synthesizers and samplers. Unlike audio recording, MIDI production provides an unprecedented amount of scope over both the performance and sound of the music, although it can sometimes lack the life and energy of music performed by real musicians.

Of course, it's highly likely that most projects in Logic Pro will use a combination of both audio and MIDI recordings together to create the optimal presentation of your track. In that respect, the combined power and functionality of the Tracks area in handling such projects makes Logic a superb production solution.

2.3 Logic Pro Main Window and the Tracks Area

Logic Pro's Main Window is the nerve center of your workflow, providing a range of different editing and arrangement features to piece your project together. Placed in the middle of the interface, and covering the majority of the screen, is the Tracks area, displaying a list of current tracks and instruments residing in the project (in the track list down the left-hand side of the Tracks area), alongside the various regions that form the structure and development of your song over time. By adding or deleting tracks or moving regions around the screen, we can visually build up the structure and arrangement of the song accordingly.

Figure 2.1 The Tracks area provides a visual overview of the tracks used in the project, as well as the arrangement of the song defined by the various regions.

KNOWLEDGEBASE

What Is a Region?

One of the principal concepts of a DAW system like Logic Pro is the notion of the nondestructive edit. Unlike the process of editing tape – which physically cuts the tape into two – an edit in Logic simply changes the proportion of data played. In that respect, we need to make an important distinction between the "source file" or recording and a "region." In effect, a region is simply a window on the original data, playing back either a small or large part of file. You can also have a number of different regions focusing on different parts of the original file, but, ultimately, all sourced from the same material. Of course, all of these types of editing would be impossible to achieve with a physical, destructive editing medium like tape.

Figure 2.2 A region is simply a window on the original data, allowing you to resize the edit at any point, effectively bringing back audio or MIDI data after the original start or end point has been established.

Figure 2.3 Use the other areas to the side of the Main Window to access more detailed application-specific editing tools and features.

Springing up from the sides of the Logic Pro main window are a number of other functional "areas." These areas relate to activities involved in the production process and the tools of editing, arranging, and mixing audio and MIDI data. For example, in the initial stages of production, you might need to input the various signals to be recorded, checking their respective recording levels and making a suitable "monitor mix" – in this case, opening up the Mixer at the bottom of the Tracks area. Later on, you might want to import some additional audio files – using Project Audio Browser – or edit the timing of a MIDI performance using the Piano Roll.

The idea behind the Logic Pro main window, therefore, is that it can provide a dynamic working environment, optimized for the tasks you need to carry out at any point in time. Although it's possible to open and shut many of these extra areas using on screen tabs, or double-clicking on regions, it's well worth memorizing some of the important keyboard shortcuts so that you open them "on the fly." Importantly though, whatever additional area you decide to open up, you'll still have some overview of the project's arrangement, allowing you to keep a handle on how your edits affect the entirety of the track.

2.4   Editor Areas – Mixer, Audio Track Editor, Piano Roll Editor, Score Editor, and Step Editor

Opening from the bottom of the Logic Pro main window, the various editor areas within Logic allow you to work with and edit your project in a number of different ways. You can open the editors using either the keyboard shortcuts described below or the corresponding icons found in the top left-hand corner of the control bar.

Figure 2.4 Use the icons at the top left-hand corner of the control bar to open the various editors in Logic.

Mixer (Keyboard Shortcut – X)

The Mixer balances the respective levels of the tracks included in your project, as well as instantiating different signal processing plug-ins (like reverb, compression, or equalization) to process each mixer channel. In addition to audio mixing, the Mixer is also the primary port of call for using virtual instruments (like synthesizers, samplers, and vintage keyboards), allowing us to create music completely within the realms of Logic. Alongside the Tracks area, the Mixer is undoubtedly one of the most important day-to-day production tools used in Logic.

Piano Roll Editor (Keyboard Shortcut – P)

The Piano Roll Editor is the main MIDI editor in Logic Pro, used to edit MIDI information that in turn controls virtual instruments or external hardware synthesizers and samplers. Using an intuitive display

Figure 2.5 The Mixer allows us to balance the mix, process tracks using audio effect plug-ins, and instantiate virtual instruments.

of note information based on the position on the piano keyboard (on the vertical axis) and the horizontal axis displaying both the note's starting point and duration, Piano Roll Editor offers both unparalleled precision and ease-of-use for MIDI editing. Using the extended editing functions of the Piano Roll Editor, you can also transform and edit MIDI information in ways that would be extremely time-consuming (if not impossible) to perform by hand.

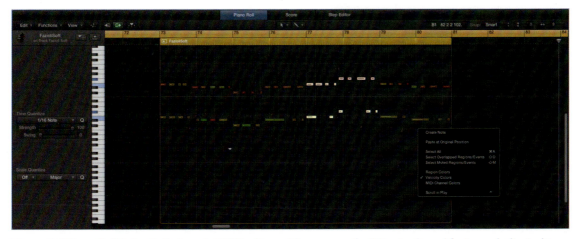

Figure 2.6 Piano Roll is Logic Pro's main MIDI editor, with a range of editing tools and features dedicated to the precise editing of performance data.

Audio Track Editor (Keyboard Shortcut – E)

The Audio Track Editor provides detailed, sample-accurate audio editing of a given region in the Tracks area. Although a large amount of audio editing can be achieved in the Tracks area alone, the Audio Track Editor lets you explore one level deeper, especially with respect to Flex Pitch and Flex Time editing, where you can manipulate the pitch and time-based properties of your audio – whether to fix poor intonation on a vocal line, for example, or quantize a drum loop.

Figure 2.7 The Audio Track Editor facilitates sample-accurate audio editing as well as accessing extended features like Flex Pitch and Flex Time.

Score Editor (Keyboard Shortcut – N)

In addition to providing a useful way of visualizing and editing MIDI information for musically literate users, the Score Editor is the vital conduit between your raw MIDI recordings and a finished printed score readable by real musicians! Although not everybody will want to use the Score Editor, it does illustrate Logic Pro's intention to be a versatile and competent music production tool in the widest sense.

Step Editor

This final form of MIDI editor offers a more specialized solution to working with controller data – like volume or pan – as well as offering an alternative means of programming and creating drum tracks. Information is presented within a unique series of vertical beams, allowing you to quickly draw controller sweeps, for example, or complicated drum patterns full of sixteenth notes. As with the Score Editor, the Step Editor might not be to everyone's tastes (there are alternative ways of achieving many of its main applications), but it does form an interesting alternative for working with several different lanes of controller data outside the Piano Roll editor.

Figure 2.8 Although slightly less versatile as a MIDI editor, the Score Editor is still a useful tool for turning your MIDI performances into a finished score.

Figure 2.9 Hyper Editor displays its information as a series of unique vertical beams. As such, the Hyper Editor forms a useful solution for dealing with MIDI controller data.

To change the respective proportion of the screen devoted to the main Tracks area and the various editors, click and drag on the thin gray line between the Tracks area and bottom editor area. This allows you to optimize the view based on your current priority – for example, expanding the Mixer when you need full access to its compression and equalization settings or using a minimized Mixer (with just the main fader positions viewable) when you want to quickly rebalance the mix in relation to the arrangement.

Figure 2.10 Change the relative window size (using the thin gray line between the areas) to prioritize the view based on your current production activity.

2.5   The Left-Hand Side Part 1 – The Library (Keyboard Shortcut – Y)

As you should be well aware of, Logic Pro X comes with a diverse sound library, including a range of playable instruments that can all be accessed via the Library on the left-hand side of the interface. Once you've opened the library you'll be able to browse a series of patches based on the current selected channel strip. If you're resting on an instrument track, for example, you'll be able to browse a collection of playable virtual instruments, covering everything from piano sounds through to virtual drum machines. If you're resting on an audio track, however, you'll be able to load preset compression, EQ and FX settings that compliment that particular instrument.

When you load a Library patch be aware that Logic might instantiate several channels worth of information, as well as FX busses unique to each instrument. In short, the Library can soon fill-out a project with a large number of channel strips and plug-ins, which can be helpful, but can also make your working environment quickly cluttered. As an alternative, therefore, you can always consider loading individual virtual instruments and effects plug-ins one at a time, and then browsing the preset settings found within the instrument/plug-in.

Figure 2.11
Use the Library, on the left-hand side of the Tracks area, to access Logic's integral sound and instrument content.

Later on, we'll see how the Library also works with the Track Stacks feature, which is an important tool for dealing with multiple track lanes that need to be grouped together. In this example, the Library also becomes an important place for you to store your own settings, which you can then move between projects, for example, or use a means of speeding up your workflow.

2.6 The Left-Hand Side Part 2 – Inspector (Keyboard Shortcut – I)

The Inspector area – found toward the left-hand side of Logic Pro's Tracks area – provides a quick-and-easy access point to key parameters relating to the regions and tracks used within your project. The Inspector is a powerful tool for audio production work in Logic Pro, performing many essential tasks like quantizing on MIDI regions or fades and crossfades with audio regions, as well as being an excellent way of viewing smaller parts of the audio mixer.

The Inspector is divided into three principal areas, which, looked at from top to bottom, are the region parameters, track parameters, and the channel strip (see below for more information). It is possible for both the region and track parameters to be minimized, in which case you'll need to open them (using the small arrow) to access the complete parameter set.

Region Parameters

Region parameters govern the playback data of a particular audio or MIDI region in the arrangement. For example, a MIDI region can be quantized, transposed, delayed, and looped, all directly from the region parameters box. An audio region, on the other hand, has its own unique set of parameters, largely in relation to the use of fades and crossfades. Besides being applied on a region-by-region basis, the region parameters box is also a powerful tool for modifying playback data across several regions at the same time – maybe applying a massed quantizing setting, for example, for all the MIDI regions in your arrangement.

Track Parameters

Track parameters establish a number of parameters in relation to the track. Audio tracks tend to have slightly fewer parameters, whereas MIDI and virtual instrument tracks present slightly more control.

Channel Strip

Not a full mixer as such, but the Inspector's channel strip view at least provides some basic access to the audio mixer on a track-by-track basis. The second channel strip illustrates the "next step" in the channel's signal path. For example, this could be the main outputs, or an aux fader, if the track was being fed to reverb or delay plug-in via a bus send.

Figure 2.12 The Inspector provides a quick access point to parameters relating to audio regions and tracks, as well as a track-by-track display of the audio mixer.

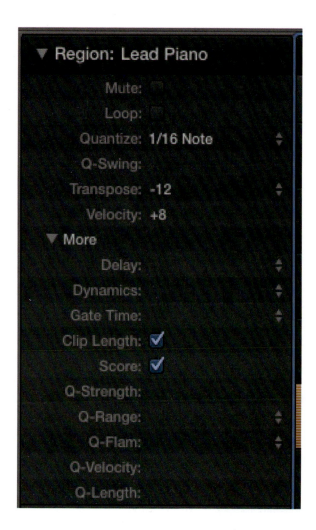

Figure 2.13
Each region has its own parameters, which can be viewed and edited via the Inspector.

2.7 The Right-Hand Side – Browsers, Loops, List Editors, and Project Notes

The right-hand side of Logic Pro's main window deals with a variety of list-based information, covering everything from a list of all the audio files contained within a project, through a list of tempo changes. All these facets are important to the day-to-day running of Logic, as well as fine-tuning.

Browsers (Keyboard Shortcut – F)

The Browsers area is a window on all the file-based activities involved in creating a project in Logic Pro. To make things easier, the Browsers area is divided into three key sections – Project Audio Browser,

Media, and All Files. Arguably the most important part of the Browsers area is the Project Audio Browser, which contains a complete list of both the audio files and the audio regions contained within a given project. In the Project Audio Browser you can import new audio files, for example, delete or rename existing files, as well as give your project a spring clean.

Use the Media Browser in situations where you want to import songs from your iTunes folder or movies from your Movies folder. Finally, the All Files Browser is a great way of negotiating a range of files on

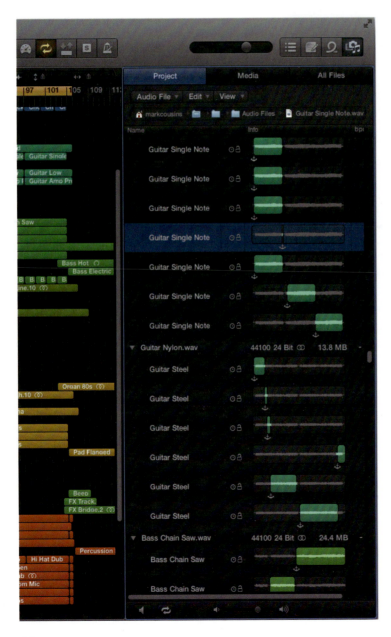

Figure 2.14
Use the Project Audio Browser to see the range of audio files and regions attached to a project.

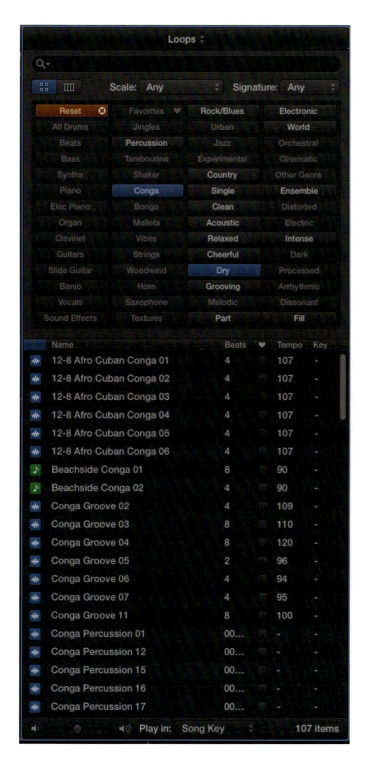

your hard drive, much like Finder, with the added benefit that you can also use it to import individual tracks, plug-ins settings, and media files from other Logic Pro Projects.

Apple Loops Browser (Keyboard Shortcut – O)

Apple Loops are a range of prerecorded musical segments that are installed alongside Logic Pro X and should be immediately familiar to anyone who's used GarageBand. Use the Apple Loops Browser as a means of both auditioning and importing these loops into your Logic Project.

Project Notes/Track Notes (Keyboard Shortcut – Alt + N)

The project/track notes features is a handy way of keeping track of information that needs to accompany either a project as a whole or the tracks within it. Use the Project Notes feature to keep a "things to do list" alongside your song, while the track notes can be useful as a means of storing recording information like the choice of microphone used and any effects written into the audio file.

List Editors (Keyboard Shortcut – D)

Although not as glamorous or intuitive as some of the other editors, these List Editors are still vital to the success

Figure 2.15
Use the Apple Loops Browser to select and import Apple Loops that you want to include in your project.

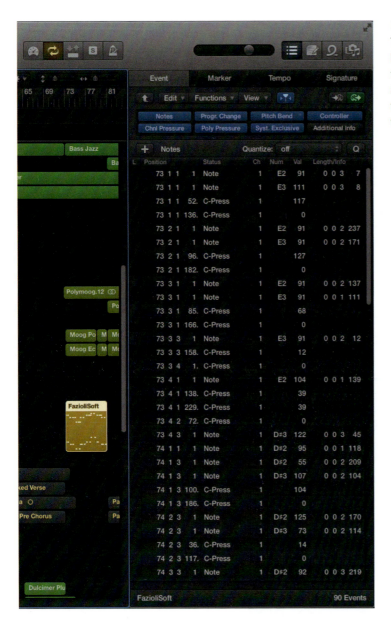

Figure 2.16
The Lists area manages the various list-based processes in Logic – from the detailed viewing and editing of MIDI information in the Event list through to the Tempo list's set of tempo changes.

of the project and the flexibility and precision of working with Logic. In total, there are four types of list available. Event presents MIDI information in its "raw" form, allowing you to see precise parameter values and positional data. Marker displays a list of locational reference points (like verse, chorus, middle eight, and so on), which can be an effective way of navigating long projects. The Tempo list manages a project's tempo changes, particularly vital when working to picture, while Signature offers a similar control in relation to time signature.

2.8 Control Bar and Toolbar

Control Bar

At the top of Logic Pro's main window you find the Control Bar. The first and most important part of the control bar is the transport controls, which can engage Logic into playback, recording, as well as being used to navigate the project's timeline. Working in tandem with the transport controls is the LCD area, which displays the current position in the song (in bar and beats and timecode format) as well as aspects like the tempo and time signature.

To the right of LCD area is the Modes and Functions buttons covering areas like the ability to cycle around a given point in the song, whether the recording mode is in overdub or replace, and of course, the all-important metronome. The remaining buttons on the control bar govern the various views and editors that can be called into action. On the left-hand side, for example, we can see icons for the Library, Editors, and the Mixer discussed earlier. On the right-hand side of the Control bar you find buttons for the various Browsers and List Editors.

Figure 2.17 Use the toolbar to access features only usually available via keyboard shortcuts or the menu.

Toolbar

Popping out from underneath the Control Bar is the Toolbar, which provides a number of handy buttons in relation to working with regions in the Tracks areas, among other things. Many of the functions with the Toolbar can be initiated with keyboard shortcuts; hence, to save space on your display, it might be beneficial to close the Toolbar. However, if you're new to working in Logic, these accessible buttons are a good way of understanding the variety of editing techniques you can apply.

To match your preferred way of working, it's worth noting that both the Control Bar and the Toolbar can be redesigned as you see fit by Ctrl-clicking on part of their grayed-out surface. In both bars, you can add further functions or remove features that aren't so important to you.

Figure 2.18 Navigate your project and invoke various Mode buttons via Logic Pro's Control Bar.

2.9 Tools, Local Menus, and Contextual Menus

With many of Logic Pro's principal audio and MIDI editors, alongside the main Tracks area, you'll find a series of tools dedicated to carrying out different activities or tasks. These tools can be configured for

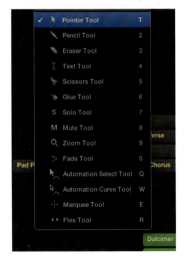

the left- and right-hand mouse button using the small drop-down menus in top center of the respective area. For example, the Tracks area includes tools that relate to the manipulation of regions, as well as automation data. When you're editing automation data in the Tracks area, therefore, it might be beneficial to place the main pointer tool on one mouse button, with the Automation Curve tool on another.

Pressing the keyboard shortcut T at any point will also bring up the particular toolbox menu for that area, allowing you to quickly switch tools without having to travel to the extremes of the area's window.

Alongside Logic's global menus, you'll also find a set of local menus unique to each area, many of which contain powerful editing features and functions that go beyond what can be achieved by mouse alone. Although not essential, it's worth familiarizing yourself with these menu features over time.

Figure 2.19 Each area has its own unique set of tools, which can be quickly accessed via the keyboard shortcut T.

In many cases, it is also possible to access these features via the use of contextual menus, directly activated by Ctrl-clicking in the relevant area. As you'd expect, the precise role and content of the contextual menus changes between areas, but you'll also find that the menu adapts given the precise part of the screen or objects you click on. For example, in the Tracks area, the contextual menu will differentiate between Ctrl-clicking in the main region area, regions themselves, the bar ruler, and the track list, with different functions appearing each time. Although bewildering at first, as you get more used to these features, you'll find the contextual menus a powerful and immediate means of accessing the real depths of Logic's editing prowess.

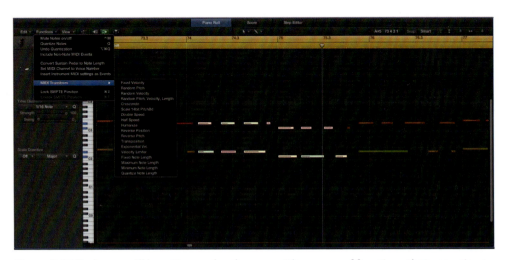

Figure 2.20 Each area will have its own local menu, with a range of functions that are unique to the activity taking place.

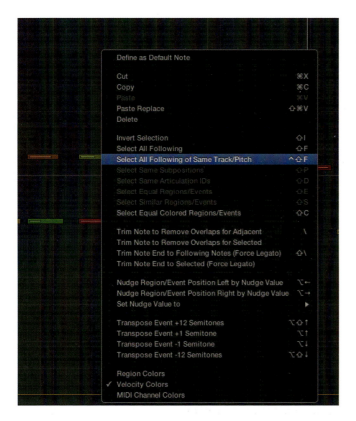

Figure 2.21
Contextual menus – activated by Ctrl-clicking on the screen – access many features that are unique to the area or object you click on.

2.10 Adjusting How You View the Arrangement: Zooming In and Out

Varying the magnification of your project is essential in terms of both keeping an overview of the "macro" arrangement of your track and to be able to hone in on certain details, even down to sample-level accuracy. In the Tracks area, the two zoom controls (for the horizontal and vertical axes, respectively) can be found toward the top right-hand corner of the window – note that you can also use Cmd+ any of the arrow keys to achieve the same movement in and out of magnification. You can also use the zoom tool to drag-enclose a specific portion of the screen you want to view, with a double-click taking you back to your original magnification level.

As an alternative to zooming in and out, you can also use the menu option View > Auto Track Zoom (keyboard shortcut Ctrl + Z) to enable a track-dependent horizontal zoom. This allows you to better view the track that you're currently working on, although, of course, this is only in relation to the horizontal rather than vertical axis.

Another type of magnification is the waveform intensity. This effectively changes the volume scaling of the waveform display, allowing you to better see quieter, more discrete parts of the waveform that you usually wouldn't be able to see. To expand the waveform, simply click and hold on the waveform icon next to the horizontal/vertical zoom controls, and use the slider to scale accordingly. Although the waveform might end up looking distorted, the meters should, in fact, confirm that it isn't.

Figure 2.22 Using the various zoom functions allows you to understand both the overall structure of your arrangement and the smallest details within it.

Figure 2.23 Changing the scaling of the waveform will allow you to improve your audio editing in relation to quieter parts of the signal.

In this chapter

Knowledgebases

Getting Connected

3.1 Introduction

Although it's easy enough to produce music solely using your computer, it's inevitable that you'll want to add some additional equipment to enhance your workflow: whether it's a USB controller keyboard to manipulate one of Logic's many virtual instruments, an additional hard drive for storing samples, or an audio interface so that you can record an electric guitar. Understanding the various connections in and out of your computer is therefore vital to getting the best from Logic and will allow you to expand your setup from a basic home computer to a full-sized professional studio!

In this chapter, we're going to take a closer look at the key components you might want to add to your existing Logic and Mac setup – understanding the pros and cons behind their selection, how they connect to your computer, and how these devices integrate into the Logic working environment. Of course, it's important to remember that no two setups will be exactly the same, so you should select the equipment that is most appropriate to your budget and the type of music you want to produce.

3.2 USB Devices

While the Mac's onboard audio facilities are respectable, they can vary from model to model. Most Mac laptops come with a microphone built-in, and again, while very respectable for speech recognition and dictation, and chatting online, it lacks the professional quality you'd want for recording music. Whatever the onboard input, you'll soon want to simply plug things in and expand your Logic studio. In this chapter, we'll look specifically at the devices that get you connected.

USB – An Introduction

Connecting devices to the Mac is an easier affair than it once was before the advent of USB. Prior to USB, different manufacturers tended to use varied, and often proprietary, connection protocols. USB brought this all together with version 1.0, essentially resulting in one connection protocol for a multitude of different uses.

USB 1.0 came as a slow protocol of 12 megabits per second (Mb/s), which was ample for devices such as mice, keyboards, printers, MIDI interfaces, and so on. Early Macs and some devices you will come into contact with (such as some older audio interfaces) are still USB 1.0. However, USB 1.0 would slow quickly with too much traffic via multiple connections using a USB hub.

FireWire was readily available at the same time, and being 400 Mb/s, it handled many of the heavier duty hardware such as external hard drives and larger audio interfaces. FireWire is still the preferred choice for many audio professionals, with the protocol now boasting 800 Mbs/s.

Fortunately, USB 2.0 came along around the year 2000 and offered a considerably faster connection speed of 480 Mb/s allowing the USB to be fast enough for hard drives and audio interfaces with higher track counts and higher data rates.

USB 3.0 was formally ratified in 2008 and has gained quite a slow prominence. This has a data rate up to 10 times faster than USB 2.0.

In practice, most external devices and peripherals for computers currently are connected using USB, including many used for music making. USB is a flexible system that enables multiple devices to be connected to one computer with the use of a USB hub. Each hub acts as a splitter/merger box and typically allows for between 4 and 10 devices to be connected at once. The USB limit is much larger than this, but is unlikely to be exceeded by most users. Let's take a look at the types of USB devices we can use with Logic Pro X.

USB Audio Interfaces

Audio interfaces come in many forms, but typically, USB-based devices are self-powered and historically (because of USB 1.0) handle two inputs and two outputs. In some instances, there may be useful MIDI sockets found here too, which we'll discuss in a little detail later in the chapter. These usually inexpensive units may not always offer the highest grade sound quality, lowest latency (see Knowledgebase 1), or the largest range of sample frequency and bit-depth options, but they are excellent solutions for where basic input and output (I/O) and portability are the key. Ideal uses for this type of interface are simple stereo-pair recordings and overdubs. A large-scale live recording would require a larger interface with a comprehensive I/O count, but to track up, layer-by-layer, a large work could be easily built upon with this type of interface using overdub techniques.

Figure 3.1
The M-Audio M-Track Plus is an inexpensive audio interface with an impressive number of features for musicians on the move. Image courtesy of M-Audio.

There are a huge variety of USB audio interfaces available, and choosing the right one can be difficult. Interfaces can vary widely in price, depending on sound quality and features. Manufacturers such as M-Audio, Edirol, Cakewalk, and others make a wide range of excellent inexpensive interfaces for this purpose, such as entry-level M-Audio M-Track Plus, shown above. This unit is well built and has lots of features such as inputs for different types of microphones (including phantom power), electric guitars, and other connections. This device can handle sample rates of up to 48kHz and bit depths of up to 24. Additionally, it boasts both SPDIF and MIDI inputs and outputs, making this a great portable starter.

Apogee, a manufacturer covered in more detail later in this chapter, has introduced a number of audio interface solutions for the Mac specifically and latterly for iPad. Their strategy has been to bring professional quality audio to these devices. The ONE (the illustrated version is the ONE for iPad and Mac), which is a small USB interface, has a microphone built-in plus an input for an instrument or a favorite microphone. The ONE comes with the professional quality output you'd expect from Apogee.

The key features vary from device to device; hence, choosing the right device for you will be dependent on the type of music you make to some extent. If you mainly produce electronic music, it is likely that a two-in and two-out device will be ample. However, if you are in a band and want to track your next album of live performances, you'll require something with more inputs and flexibility. Many devices can handle multiple microphone inputs and also include some additional features such as software mixing and digital signal processing such as compression and effects.

USB version 1.0, unlikely to be adopted on many modern devices, was a very slow connection, and many of the stereo interfaces were restricted to 44.1 kHz and 16 bits. Most devices are now USB 2.0 and adopt the data rate of 480 Mb/s, which is a little faster than that of FireWire 400. Therefore, while in theory any USB 2.0 audio interface could handle a large I/O count, there would be interruptions for other devices connected to the computer using the shared USB bus (mice, keyboards, printers, etc.); thus, it is sometimes considered a little less reliable. This is due to the data throughput originally not being as stable as FireWire. As such, FireWire tends to be adopted for larger audio requirements or even PCI Express (PCIe), which are both covered later.

Figure 3.2
Apogee has teamed up with Apple to produce some excellent audio devices. Illustrated here is the new iPad and Mac compatible ONE, which has a mono input of a built-in microphone, jack input, or external microphone. Image courtesy of Apogee Electronics Corporation.

Figure 3.3
RME's Total MIX FX is an excellent example of the additional mixing features that can be provided specifically for RME's audio interfaces. Image courtesy of RME.

USB Controller Keyboards

To play in notes through Logic's instruments, you will use the Musical Typing Keyboard provided by pressing Cmd + K on your Mac, but sooner or later, you'll want to escalate to your own controller keyboard. Many keyboards come with MIDI sockets (see the MIDI section toward the end of this chapter) to connect to your computer, but recently, these keyboards have connected directly using USB.

Figure 3.4
Pressing Cmd + K brings forward the Musical Typing Keyboard that allows the Mac's qwerty keyboard to become a useful tool for entering in music on the move.

The introduction of a USB controller keyboard negates the need for a dedicated MIDI interface if you mainly use the internal Logic Instruments. However, if you wish to use some older MIDI equipment, you'll need to use a MIDI interface, and for more details on this visit page 53. M-Audio's Oxygen 25 shown on the next page is a good example of a portable and flexible controller keyboard offering both two octaves of input in addition to some assignable controller knobs. With a keyboard like this, you can enter in the notes and also control some of the parameters of the Logic Instrument such as the filter and envelope settings (discussed later in this book).

Figure 3.5
M-Audio's Oxygen25 is a common and practical USB keyboard for musicians on the move, offering both a couple of octaves and transport and mappable controllers.

Figure 3.6
M-Audio Axiom 49. A four-octave keyboard with a whole host of controller controls and drum pads for music creation.

Many controller keyboards are available across the product ranges, working up to some pretty hefty weighted numbers. The top-of-the-range controllers are often built to feel like a real piano's action under your fingers and are solid workhorses. There are many examples in between such as the M-Audio Axiom 49, which additionally includes controllers and drum pads.

Select your controller carefully depending on the music you create and the flexibility you need. Dance music production would probably benefit from a controller keyboard with assignable knobs and sliders to various synthesizer and sampler parameters, while a virtuoso keyboard singer/songwriter might prefer to ensure the feel of the keys are as close to a real piano as possible. Portability will be a key aspect, and this may heavily influence your choice.

3.3 FireWire Devices

FireWire audio devices have long been the standard for professionals on the move. The reason for this was due to the aforementioned speed and reliability issues of USB 1.0. As such, FireWire became traditionally used for the data hungry devices such as audio interfaces and hard drives. FireWire still remains popular as its ports are rarely shared with many other devices that could interrupt the data flow. However, Thunderbolt is gaining prominence and shall eventually supersede FireWire we suspect.

Meanwhile FireWire audio interfaces remain popular, ranging from the inexpensive Focusrite Pro14 (shown below) with four analog inputs and up to six outputs (four analog and two digital) to more expensive units with a higher count of inputs or outputs.

Figure 3.7 Focusrite's Saffire Pro 14 is part of a series of FireWire-enabled audio interfaces that, with Apple's FireWire converter, are fully compatible with Thunderbolt. Image courtesy of Focusrite.

The latter devices are ideal for the larger projects you might wish to get involved in. Mid-range interfaces often offer somewhere between four and eight microphone inputs and up to 10 or sometimes more analog outputs. In addition to this, there are often a multitude of digital connections, such as the ADAT lightpipe standard, which can add another eight inputs and outputs to the interface with a suitable convertor. This method of expansion is typical of many audio interfaces, and if paired with a digitally connected console or multichannel convertor unit, such as Focusrite's OctoPre MkII, most of them can be made of the units full input and output count.

Apogee, introduced earlier, is a well-respected, high-end audio converter manufacturer who has in the past worked closely with Apple. A result of this cooperation saw the introduction of a number of professional facing audio devices. These Macintosh-only audio interfaces contained some excellent and specific benefits for Logic users. From the start, these units boasted Apogee's reputed high-quality converters, which after all are a key part to the sound quality of your studio setup among many other key features.

The first that was introduced was the Ensemble, now sadly discontinued, that boasted eight analog outputs and eight audio inputs, four of which are very low-noise, digitally controlled microphone preamps. This was a well-received device.

The smaller and more portable Duet, still in production, is essentially a two-input/two-output portable device. The accompanying breakout cable allows for six connections: two XLR microphone inputs, two jack instrument inputs, and two line-level outputs. The Duet also comes with Apogee's Maestro software that interfaces between Mac OS X's Core Audio and their hardware. This is comprehensive and offers the user the ability to configure multiple headphone mixes and outputs plus many clocking options and other intricate and useful controls. The software can digitally control the microphone preamplifiers by mouse using the software, but ingeniously, Apogee has also included a knob on all their interfaces, which they call a "precision digital encoder" for the purposes of controlling faders and knobs on screen.

Figure 3.8 From Focusrite larger rack-mountable devices such as the Liquid Saffire Pro offer a considerable number of inputs and outputs. Images courtesy of Focusrite. Also illustrated is Apple's Thunderbolt to FireWire convertor, which is fully compatible with all Focusrite FireWire interfaces.

Figure 3.9

Naturally, there are a plethora of FireWire audio devices on the market for use with Logic, which work with Apple's Core Audio. However, before taking the plunge, it might be worth checking on the manufacturer's website that the device you are considering is compatible with Core Audio and Logic.

3.4   Thunderbolt

In recent years, the Thunderbolt protocol has started to take further prominence on Apple portable devices. As Apple's engineers worked ever harder on the miniaturization of laptops and mainstream Mac computers, solutions were needed to overcome the physical size of the plugs on the side, such as the VGA output, FireWire 800 (which is larger than the older 400 plug) and Ethernet. Working on the same hub philosophy as that of USB, and less so FireWire, Thunderbolt could probably replace high data connectivity connections such as USB, but is most likely to be used to offer a replacement for FireWire and also users wishing to continue reaping the benefits of their PCI cards. For newer Macs, PCIe slots are no longer offered and alternatives are needed, via Thunderbolt, to keep these systems running.

Look out for some exciting peripherals with Thunderbolt on as time goes on.

Figure 3.10
Apogee's
Thunderbridge
is a useful unit to
connect older large
Apogee gear to the
new Thunderbolt
protocol.

3.5   PCI Express

Perhaps you have an existing studio with an analog console or are replacing another multitrack system with Logic. Either way, to obtain a truly large set of inputs and outputs, a top-flight system will be required using an earlier generation Mac Pro's PCIe slots.

These types of systems bypass the FireWire and USB connections and instead connect either directly to a PCIe circuit board that sits within your computer or via an external chassis hosting your cards and connected to the computer via Thunderbolt. PCIe contains benefits as the cards connect directly to the computer's central processing unit (CPU). This enables much faster data rates than FireWire or USB 2.0, therefore offering larger track counts, higher sample rates, and larger word counts.

There are a few good examples of this model. One is the Mark of the Unicorn (MOTU) HD192 system, which can be expanded to accept 48 inputs and 48 outputs concurrently and works with a range of software.

Figure 3.11
MOTU's HD192 Series connects to the Mac using the PCI connections and offers high sample rates and is highly expandable. Image courtesy of MOTU.

3.6 Other Audio Equipment

As you expand your set up, you'll need to consider other things that connect to your audio interface. These devices are typically microphones, loudspeakers (called monitors in professional circles), headphones, and an ancillary outboard. In this section, we'll look at these in detail.

Microphones

There are many microphones in the market, and some are indeed collectors' items. However, you'll need to know a few simple facts to find the right choice for you and get them connected. Essentially, there are two common types of microphone, different in the way in which they work internally: the dynamic microphone and the condenser microphone.

Both condenser and dynamic microphones have their own unique benefits, and this can be used to great effect depending on what you're recording.

Condenser microphones are very sensitive, being able to capture really quiet sounds as well as handling loud ones too. Condenser microphones can also capture a wider range of frequencies, from very low bass through to the very top treble. The cost of these relatively fragile microphones was for many years out of reach for most people, but recently, a wide range of affordable condenser microphones have swamped the market due to Chinese production. Condenser microphones require a form of power source to operate. Typically this is known as phantom power, and it may be worth ensuring your audio interface can deliver this when making a purchasing decision. Phantom power sends a 48v DC signal down the microphone cable to power typically two things: the diaphragm and the internal preamplifier.

Figure 3.12
Neumann microphones have been an icon of vocal recording for some time, and the U87 shown here is a perfect example.

Electret condenser microphones differ slightly in that a charge is built in to the capsule containing the microphone diaphragm. As such, only power for the onboard preamplifier is required. These can, like traditional condenser microphones above, run with phantom power, but electret's main benefit is that they have a provision for battery power onboard when no phantom power is readily available. A classic example of this would be the AKG C1000S shown below.

Figure 3.13 The AKG C1000 is a popular example of an electret condenser microphone that has capacity for a PP3 microphone onboard to power the onboard preamplifier.

Figure 3.14 The SM58 from Shure is possibly the most common of all dynamic microphones offering legendary solid construction and performance for vocalists live.

Dynamic microphones are traditionally less expensive and thus offer users the ability to record anywhere. They can accept loud sounds well (close miked drums, guitar cabs, and the like), as well as being able to reject spill as they're not as sensitive as their condenser counterparts. This is of great use when recording drums, for example, as the dynamic mic's typical frequency range is less able to handle the highest of frequencies than a condenser. One other attribute is the dynamic's sturdiness.

Connecting each type of microphone will be typically via the XLR connections provided on your audio interface. It is worth again pointing out that condenser microphones require phantom power (+48 V), which will need to be engaged once you've connected your microphone. Ensure your speaker levels are down as this can cause quite a large bang.

Monitors

Monitors are loudspeakers to you and me. In professional terms, they allow you to "monitor" the signal, hence their name. There are literally hundreds of products on the market currently and choosing a pair of monitors that suit your music and your budget will require a little investment of time.

Studio monitors can range from expensive (and often large) far-field studio monitors through to smaller, and often cheaper, near-field monitors.

The studio monitor market has blossomed in recent years. As a result, there are many excellent units to choose from. Many monitors represent exceptional value for the money, and care needs to be taken when choosing them. For example, if you're doing Dubstep, you'll probably want something that can handle bass well.

Most monitoring systems these days are what we call "active," meaning that they come with specially developed amplifiers that work best or are tuned for that loudspeaker unit. Active speakers can represent the best performance for the money. "Passive" monitors require you to obtain an amplifier to power them, which invariably means additional expense.

Figure 3.15
Controlling volumes in studios where a mixing console is not necessary can be resolved using Logic's Master Output Fader or investing in a specific solution such as Mackie's Big Knob.

Some active units also come in what is known as a 2.1 configuration. Don't be put off by the numbers. These just relate to two monitors (left and right stereo) and the ".1" refers to a subwoofer, a speaker unit designed to deliver bass typically below 150 Hz. It is likely that you've heard of 5.1, which relates to the surround sound format commonly used at home for surround sound. Logic can allow you to mix in 5.1 (see Chapter 10). This relates to five full range speakers (left, right, center, rear left, and rear right) and the subwoofer ".1." There are increasing numbers of speakers that are added to this, such as 7.1 and 10.1, for other surround speaker applications.

Connecting active monitors to your set up should be as simple as connecting them to your audio interface. However, it is worth noting that many active monitors do not provide conventional volume controls. As such, you will need to control the monitor level from either Logic using the master fader, from your audio interface (if a level control is provided), or by some other means, such as a mixing console if you have one. Specifically designed units such as Mackie's Big Knob can offer a neat solution. The Big Knob is more than just a volume control. It offers management of headphone mixes, talkback, and routing to other devices.

Headphones

Studio headphones (also known as "cans") are another important element of the studio. They are really not very different to consumer headphones that you use already at home. Often, studios will provide what are known as closed headphones, meaning that sound from the outside is rejected as much as possible and vice versa.

Recording acoustic music without good closed headphones can lead to either the headphone mix spilling out and being heard in the vocal recording or the loud drums or guitar stack you're recording being heard over the headphone mix.

Additionally, how loud they operate can be an issue if you're a drummer and want to hear the rest of the band in the cans, but good closed designs should prevent the need for too high a level.

A good, solid, and reliable example of a studio set of headphones is the Beyer Dynamic DT100 shown left, which has been used in studios across the world for over many decades.

Figure 3.16 The DT100 headphones from Beyer Dynamic have an impressive history spanning decades. The timeless appeal is due to its closed design and user-replaceable parts.

Outboard

From time to time, you may wish to add an analog processor such as a favorite compressor or effects unit. Doing so in an analog studio is a simple matter of patching it in, but integrating one into Logic will require some further consideration.

For example, if you wish to record using your favored compressor as an insert, you may wish to see if your audio interface can accept inserts. Otherwise, connecting this device will be via the inputs and outputs of your interface, allowing you the opportunity to route the signal through the processor and back into the computer again. This may cause delay in the form of latency unless you're able to use onboard routing software similar to Apogee's Maestro software.

These inputs and outputs could also be employed as an auxiliary send and return in the case of an external effects unit. Latency should not cause too much of a problem in such instances as this may be masked dependent on the effect employed.

The expansion of your studio in this way will depend on the amount of analog inputs and outputs you have at your disposal.

3.7   Integration into Logic – Audio Preferences

Apple's operating system contains a neat feature called Core Audio. Core Audio should mean that when you launch Logic, each connected audio interface appears in the audio preferences. In fact, Core Audio and Logic should, theoretically, allow for an interface to be connected to the system mid-session, or "plug-and-play" as it is known. Logic alerts you that a new interface has been added and asks you whether you would like to use it, as in the example below.

To check your interfaces and settings, simply go to Logic Pro X > Preferences > Audio, where the Devices tab should be present. Here you'll be faced with the information relating to Core Audio connectivity. Those used to other features in previous versions of Logic will remember the ability to connect to a range of ProTools interfaces. This feature is no longer available in Logic Pro X. Connecting Avid ProTools hardware will be via Avid's own Core Audio Manager software. This typically limits the inputs and outputs to their device. More information can be gleaned from www.avid.com.

On the Devices tab there are many parameters that can be altered to get the best from your audio interface(s), which are called Devices in Logic. Here you can select the Core Audio-compatible device for your Logic system at the time. Simply select the one you wish to use, and presto, you should be up and running. Core Audio makes it that simple.

Getting the very best from your device may mean managing the latency you experience. As we discuss in the Latency and Monitoring Knowledgebase, there are workarounds regarding monitoring. However, the computer's driver and interface will also require time to "buffer" the information together before sending it out. The I/O buffer size is measured in samples, and this therefore relates to the amount of time the computer and device take before you hear the sound. In an ideal world, we'd want this buffer to be at its lowest setting of 32 samples, which at 44.1 kHz would be considerably less than a millisecond and a perfectly acceptable delay. Doing this will require a very powerful computer, as more of the processing power is needed and may result in your audio dropping out or distorting. The ideal is to find a happy medium where the delay is not too long and the computer's processing power enables unhindered audio. If latency causes problems with your recordings, you can use the Recording Delay feature to move the audio back or forth so that it fits with the rest of your music. The default for this is 0 ms but can be moved by 5000 ms either way.

Figure 3.17
When a new audio interface is added, Logic will notify you and ask you whether you want to use this device. If this does not appear, then it might be worth obtaining the latest drivers for the interface from the maker's website.

Bit rates in digital audio refer to how fine the measurement of the amplitude, or loudness, of the audio signal is. CD recordings are presented at 16 bits and were considered resolute enough for the domestic market. However, within music production, 16 bits can be considered a little too coarse to the ear and as such 24 bits is now desirable, as it measures the amplitude scale in finer detail. Therefore, by increasing the detail and accuracy of the waveform, it is more representative of the original audio. To record using 24 bits, select the 24-bit Recording check box within this pane. Remember, though, that your audio interface will need to respond to 24 bits to benefit.

Higher bit rates than 16 bits are desirable, but come at a slight price. The data this occupies on your disk are larger than that of plain 16 bits and as such you should consider whether you have enough hard drive space before the session (discussed in this chapter). Roughly, you'll need nearly twice as much hard drive space as with 16 bits.

Input monitoring only for focused track and record-enabled tracks' check box acts much like the old "input monitor" button on a multitrack tape machine in so much as those tracks that you record-enable become audible rather than the signal from the tape itself. Using this feature now means that it is automatically enabled rather than expecting you to use the "I" button on the track headers and will open the focused track or record-enabled track.

Figure 3.18 The "I" button shown on Audio 1's track header here permits you to switch to input monitoring rather than listening 'off tape' as it were. The Audio Preferences pane allows you to permanently select Input Monitoring for focused tracks.

Another key checkbox selection here is Input Monitoring Level for Record-Enabled Channel Strips. This essentially allows the fader on the channel to work as a level for input monitoring, which will not affect the level at which you're recording.

The Process Threads selection menu offers users the ability to override the automatic number of audio processing threads available to Logic. Users should, as the manual advises, keep this setting to automatic, but you may wish to experiment if using more audio tracks.

The Process Buffer Range refers to the amount of time you will allow the computer to gather and process information into the buffer before playing it. The lower the range, the faster the computer will respond and the lower the latency. However, the computer may be unable to process all the required information in the time specified and drop out altogether.

ReWire, discussed later in this chapter, is like an internal audio and MIDI patch lead connection between two pieces of software such as Reason and Logic. ReWired Behavior refers to the type of work you are doing between Logic and the ReWire device and hence the burden on the computer. The choices are between playback and live operation. Playback assumes that the computer is being played and can use the buffer to its best effect to ensure timely playback of instruments, and this process requires less CPU load. However, live mode assumes that you are playing the instrument live and as such requires the connected device to be more timely when producing the sound; the computer needs to prioritize this, and therefore, more CPU load is used.

The Scrub speed and response will reflect the way in which you hear audio as you use the scrub tool. You may choose to make some edits if your editing requires a different method of operation. Experiment as you need to, but we suspect you'll keep things set as they are.

KNOWLEDGEBASE 1
Latency and Monitoring

When recording was in the analog domain, the signals fed from the mixing console through to an open-reel multitrack tape recorder and back would only be delayed by an acceptable millisecond or so. To all intents and purposes, this was perceived as pretty much instantaneous, certainly within acceptable margins. As this delay increases, performers can be put off their performances, and they might appear to be out of time.

This delay in modern digital audio is called latency and is a common side effect of working with digital audio workstations such as Logic. Latency is caused by the delay experienced as your computer, audio, and MIDI interfaces process the information. A common example of this is the delay between pressing the MIDI keyboard and the sound generated from an audio instrument in Logic.

Overcoming latency can often be tricky as it depends on the speed of your audio interface and computer. Monitoring is one major issue to overcome, and this can be achieved by employing an external mixer to handle your headphone mix. Many inexpensive USB interfaces combat latency by employing a "mix" knob on the unit, such as Digidesign's (now Avid) original MBox. This allows

(Continued)

you to blend the input signal with the output of the computer, which allows you to hear your input signal prior to any incurred latency. Some Audio Interfaces come with their own software-based audio mixer, which performs this monitoring duty, as well as often integrating onboard digital signal processing. If one of these solutions is used, then it is desirable to switch off the Software Monitoring checkbox in Preferences > Audio. Another solution is to employ Logic's Low Latency Mode.

Latency can be problematic when performing using a virtual instrument or monitoring a recording. Logic has created a neat solution called Low Latency Mode that intelligently bypasses active plug-ins on the channel you're working on, thus freeing up processing power. This allows for latency to be managed to the preset limit of delay you have chosen in the preferences. To change this limit, go to Logic Pro > Preferences > Audio > General. Turning on Low Latency Mode is easy as there is an icon directly on the transport bar. In most cases, this will alter the sound of the channel you're working on, but this is only temporary for the time this feature is engaged.

One of the best current solutions to overcome latency with Logic is to use some hardware connected via PCIe or Thunderbolt to the Mac. This should boast some of the lowest latency available at the time of writing.

3.8   Hard Drives

Recording audio is a data-intensive exercise, and often as modern track counts, sample frequencies and bit depths increase, the internal, original hard drive of your Mac may start to strain under the pressure. As such, many Logic users have become accustomed to adding external hard drives to their Mac setups.

Choosing an external storage device will depend entirely on what you wish to do. For example, if you just want to back up your system, or your audio files, then the drive can be relatively low speed and perhaps lower cost. However, should you wish to use this addition as your dedicated audio drive, then speed and capacity could both become issues worth factoring in.

Audio files can take up some considerable hard drive space, and as such a large drive is welcome to contain all your projects. Drive speed is also an issue. It is preferable for the audio files to come off the drive as fast as they went on, and as such many manufacturers, such as LaCie, provide an option for a 7200 rpm drive instead of the slower speeds such as 5400 rpm.

Although USB 2.0, rated at 480 Mb/s, is faster than FireWire 400, its structure and ability to hub mean that it can share its valuable bandwidth with other devices such as mice, keyboards, cameras, and MP3 players. As a result USB is, at times, considered to be less reliable for data efficiency, throughput, and stability. Thus, FW400 has been the long-standing de facto standard despite it not being the fastest.

However, FireWire devices currently come in two flavors: 400 Mb/s and 800 Mb/s. Most devices still utilize the common FireWire 400 (FW400) format due to its history, popularity, and reliability. Most hard drives and devices now utilize the newer FW800 format, which can offer very fast data rates and is becoming more popular for audio duties.

Hard drive manufacturers such as LaCie make a series of drives suitable for the Logic user, from portable, self-powering drives such as their Starck range through to the industry-standard d2 Quadras. The d2 hard drive series has, for many, been the professional hard drive of choice and now supports a multitude of interface formats, usually USB 2.0 (or even USB 3.0), FW400, FW800, and the new 3-Gb/s eSATA connection all on the same unit. Naturally Thunderbolt and Thunderbolt 2 are now making appearances too.

Figure 3.19
The LaCie d2 for many has become one of the industry-standard hard drives of choice, offering both reliability and flexibility (including a rack mount available from LaCie).

SATA and Internal Storage

Serial Advanced Technology Attachment (SATA) is the connection protocol between a computer and its internal hard drives. This is a very fast method of connection working at speeds of either 1.5 Gigabits per second (Gbit/s) or newer 3 Gbit/s. If working with the previous generation of MacPro computers, it is possible to connect up to three additional hard drives to that of your system disk. The new MacPro computer is a radical change and only boasts SSD storage. You'll need to invest in an external drive if you choose this computer.

For audio and multimedia, an external hard drive is an excellent way to store projects to different drives or for back up purposes. Separating out the system drive containing Mac OS X and Logic from the audio allows the drives to be dedicated to specific tasks. However, the simple overriding benefit is the speed of SATA drives for working with high data rates such as multitrack audio.

3.9 Control Surfaces

The studio was traditionally a tactile environment where physical faders, tape machines, cables, and instruments were the norm. As the digital revolution has taken its hold, so much of that studio has been pared down to the computer screen. As such, the notion of grabbing the fader for any channel to make an on-the-fly adjustment becomes a little bit more involved, usually requiring several mouse clicks perhaps with some keyboard presses too. Computing power and software development has meant that the amount of features controlled from one mouse and keyboard has grown dramatically. A common solution to this problem is to bring back the tactile aspect of the studio in the form of an external control surface.

Figure 3.20
PreSonus' FaderPort is a simple and intuitive control surface allowing adjustment of one fader at one time, plus the usual transport controls. Image courtesy of PreSonus.

Control Surfaces come in many forms from the one fader unit such as the Presonus FaderPort to the fully fledged Mackie MCU Pro and MCU Pro XT. What is common is that most come in the form of a bank of faders with some control for other aspects of the channel strip such as pan and auxiliary sends. As you navigate around the controls, Logic should also follow suit and vice versa, giving you tactile control over your mix.

Control surfaces such as Mackie's original Control Universal were connected using the normal MIDI cables, although the newer versions (shown below) and other systems, such as the FaderPort, adopt the USB protocol. Other control surface devices are connected to the computer using USB, FireWire, and on occasion, Ethernet.

Figure 3.21
Mackie's MCU Pro has been the choice of many Logic users due to its impressive integration and overall flexibility. Image Courtesy of Loud Technologies Inc.

Setting up a control surface is easy, provided they integrate with Logic natively using an existing or accompanying driver. To install a new device, go to Logic Pro > Control Surfaces > Setup …. You'll face the Setup pane that has two menus at the top: Edit and New. Select New > Install from the menus at the top, which will open another window listing all the Logic-compatible control surfaces. You are encouraged to "Add," or you can scan for attached devices.

Once you have added your controller, close this window and return to the controller setup page. Here, it is possible to select the controller from the diagram on the right-hand side and then edit any information about the device in the inspector-like space on the left-hand side. Within this space, you're able to manage many aspects and preferences, such as what MIDI ports it is connected to and what feature the fader controls within Logic.

It is possible to add more than one model of control surface. As such, Logic makes it possible to edit what the controls actually operate in Logic within the setup page. In the example of more than one controller, duplicated buttons such as play and stop could be reassigned on one unit, while the faders could naturally lead on, say from Channel 8 on one device to Channel 9 on the next control surface. To get two control surfaces working in tandem, ensure in the Control Surface Setup window that the controller pictures form a horizontal row. However, should you wish for the controllers to act independently, then place the icons in a vertical column.

The way in which the Control Surface reacts and interfaces with Logic can be comprehensively edited using the Controller Assignments pane (Logic Pro > Control Surfaces > Controller Assignments). There are two views here: Easy View and Expert View. In Easy View, each command can be seen one at a time, and it is simple to reassign a control to another duty. To do this, simply press on the part of Logic you wish to control. Next, press Learn Mode to move a control or touch a button on the controller surface to map it accordingly.

To edit these modes, choose Expert View; the area to the left of the pane describes the zone of the surface, in this example the Mackie Control Universal. The control surface itself is split into zones such as VPots, Tracks (faders), Global Views, and so on. The next table refers to the modes in which each zone can operate, and by selecting one of the modes, the Control/Parameter list will change to show you what each control does.

The Controller Assignments editor is incredibly powerful as any feature of the control surface can be remapped to another to allow you perfect control over a particular session or mix.

The iPad "Logic Remote" Control Surface App.

New to Logic Pro X is the iOS7 Logic Remote app. This is a neat application that extends your control over Logic quite considerably by offering more than most conventional control surfaces. For example, there are several ways to input musical data other than your controller keyboard, such as the chord strip interface used in the iPad's version of GarageBand. It is quite simply more than just a remote control.

On your iPad, download the Logic Remote app from the App Store and upon launching it should ask you which Mac on the local network you'd like to link with. Logic Pro will need to be launched on your Mac for this to happen. Once you've selected the correct Mac, a dialog box will appear on the computer asking for your permission to link your iPad. Go ahead. The two apps are now linked and you're ready to get controlling Logic.

Starting out on Logic Remote, you'll note the top left icon has five squares within it. This is the menu button and allows you to quickly navigate between the pages of the application. Clicking on each of the colored tabs to the bottom of each mixer channel results in Logic Remote loading up the name from Logic Pro. Moving up the channel strips on the mixer you get a fader, pan pot, mute, solo, record enable, and automation buttons. All very handy in a remote, especially if you're in the live room ready to put down your vocals the other side of the glass from your Mac. Above the channel strips is an LED style peak meter area. Above this is the common transport and application bar.

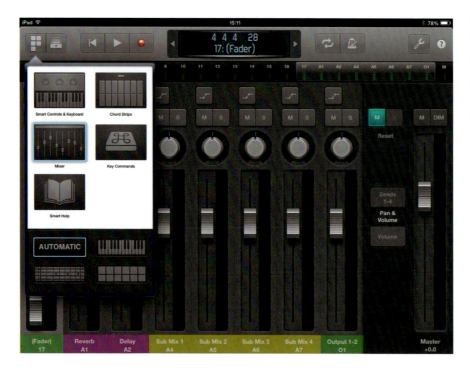

Figure 3.22
In this image it is possible to see the menu open and the five different screens offered by Logic Remote.

With a channel now selected on the mixer page, you can navigate using the top left button to another page. Choices here are Smart Controls and Keyboard, Chord Strip, Mixer, Key Commands, and Smart Help.

The Smart Controls are a useful addition to Logic Pro and these are mirrored on this page alongside a keyboard. The keyboard itself offers different sizes for those with fatter fingers. Clicking the keyboard button to the right hand side of the page above the keys provides three options to squeeze a few more notes on screen. Other controls here permit the user to control many aspects of how they play their music into Logic. You can sustain without the need for a pedal using the sustain switch, change octaves using transpose keys, and also alter how the keyboard reacts to your playing using the glissando, scroll and pitch keys. There's also an arpeggiator button which activates that favourite of effects, further enhanced by the latch switch that keeps things sustaining as you perform.

The Chord Strip page allows you to quickly interface with string based instruments, strumming chords out in a similar way to playing with the right hand of a guitar. The sustain switch is also available here and is effective in making performances a little more realistic. When you select one of the drum instruments or Drummer itself, the iPad interface alters to a visual drum kit which is usually responsive even over a mediocre network.

Figure 3.23
The Keyboard in Logic Remote is quite flexible and can be made smaller to offer a few octaves of control. Note the other controls to assist your performance.

Figure 3.24
The Chord Strip is a quick and effective way to express yourself with chords on string based instruments such as guitars.

Similarly, on the bottom of the menu page you can select whether you wish Logic Remote to choose the interface most appropriate by choosing "AUTOMATIC", or select from guitar interface, or drum pad, or fix it to be a keyboard for the duration of your session.

The Key Commands page is a very useful addition to Logic Remote as it permits a whole host of controls through sliding pages. If you keep scrolling right you'll pass through pages of controls to manage your Logic session, but finally you'll come to a page or two with blank squares. Upon touching one of these, you'll see a menu of key commands available for assignment. This feature could prove to be highly useful on specific jobs or when you're stranded in the live room away from your Mac.

The final screen in Logic Remote is the Smart Help page. This, at first, does not look to be all that interesting, but when you're deep in an editing session can be a life saver. Essentially this is Apple providing the PDF manual as part of the Logic Remote application, but it is fully searchable.

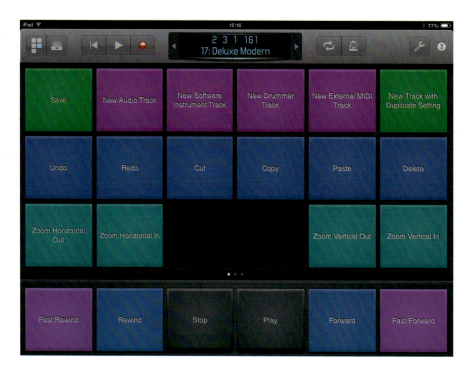

Figure 3.25
The Key Commands page is excellent, fully customizable and enables you to fully control Logic Pro from afar.

3.10 Distributed Audio Processing and External DSP Solutions

In the past and perhaps still now, there will come a time when the onboard processing power of the computer will start to strain under the pressure of plug-ins within your mix. As with Avid's HDX ProTools system that employs additional DSP chips to relieve internal processing power, there is still an external solution to the problem for Logic users too. Distributed audio processing allows for an additional DSP to be made available to the computer.

Using a system such as Universal Audio's UAD-2 system, power can be unleashed from your computer to make for a faster and more powerful workflow. The audio processing is taken care of outside the host computer using proprietary processing to Universal Audio. As such, plug-ins can be purchased to add to the system, which might be more reliant on your DSP than you would desire for your computer.

Setting up an external processing device such as the UAD is easy and is often a matter of simply connecting the FireWire and installing the accompanying driver. With OS X rebooted, Logic should now pick up that it has the system installed. Logic may rescan its audio units and an extra line within the audio units menu when selecting plug-ins called UAD. For more installation guidelines, please refer to the manufacturer's instructions and website.

SSL's Duende used to be an external system in the similar vein to that of the UAD-2, but has turned the process native, using the internal processing power of the computer to emulate the plug-ins. While this is not strictly external processing, it does illustrate that processing power is perhaps getting to a stage where external DSP may become less and less necessary.

3.11 What Is MIDI?

MIDI was originally developed as a communication protocol between hardware synthesizers, samplers, drum machines, and of course, sequencers like the original MIDI-only version of Logic. The protocol is made up of a number of different types of messages – including note on, note off, velocity, modulation, and so on – all of which can be sent to 16 available MIDI channels. In theory, the channel system was originally used to denote different devices along a "daisy chain" of MIDI leads, with each device sifting its appropriate collection of message from the collective stream of MIDI data passing along.

Nowadays, of course, hardware MIDI devices are much less common (although you may well use a MIDI controller keyboard plugged into the MIDI port of the Audio Interface), but the MIDI protocol is still alive as an effective means of "internally" communicating with virtual instruments in Logic's audio engine – a language if you like. Open up the Event list, for example, and you still get to see the list of familiar MIDI events, all with their associated channel and parameter numbers. Also, the Instrument Parameters in the inspector allow you to specify a MIDI channel for the instrument in question (although most default to ALL), and of course, instruments like the Vintage B3 use different MIDI channels to differentiate between the two different keyboard manuals and the bass pedals contained within the instrument.

So, even though you may never connect a MIDI lead again in your life, MIDI is definitely here to stay as an integral part of sequencing with Logic. It is really important to appreciate MIDI as your communication language when controlling instruments within Logic.

However, if you're itching to use an older classic MIDI synth, you'll need to connect this to your set up. As we've established, many audio interfaces come with a small MIDI interface and sockets included. However, if you do not have these or if you wish to connect a multitude of synths together, you may need to consider a dedicated MIDI interface, which connects to your Mac's USB connections.

Many manufacturers fulfill this function. MOTU is one such manufacturer, which offers a wide range starting with the Fastlane (a two-in and two-out MIDI interface) all the way through to the professional-facing MIDI Express XT (eight-in and eight-out, with many additional features, such as SMPTE timecode for striping).

Figure 3.26 The MIDI information can be clearly seen within either the separate Events List window (Windows > Events List or Apple + 7) or the Events Pane in the Arrange window by pressing "D." The Events List shows the corresponding MIDI information contained within the selected MIDI region.

To the rear of these interfaces lurk a variety of five-pin DIN sockets known as MIDI sockets. These usually come labeled in three ways: IN, OUT, and THRU. As you might imagine, MIDI IN receives instructions from another device – for example, the sound module will receive an instruction from the MIDI OUT of the Apple Mac's connected MIDI interface. If the sound module sends information back to Logic, such as MIDI controller data, then the MIDI OUT will need to be connected to the MIDI IN of the interface. These MIDI connections apply to every MIDI-compatible device within your studio.

Figure 3.27

Additionally, MIDI THRU is included in the MIDI protocol to allow for something that is known as daisy chaining. In the very early days of MIDI (pre-Logic), it was unlikely that your hardware sequencer, or Atari computer, would have more than one MIDI output, two at the very most. As such, if you had more than one MIDI device, it was important that you could connect it to the sequencer with ease. Enter the MIDI THRU socket, which replicated the MIDI IN data and passed

Figure 3.28 Mark of the Unicorn's (MOTU's) Fastlane is a solid and inexpensive USB MIDI interface, while their MIDI Express XT is packed with additional features. Images courtesy of Mark of the Unicorn.

it through the device allowing you to connect it straight to the MIDI IN of the next device unaltered. On rare occasions, some MIDI devices would merge both the MIDI OUT and MIDI THRU of a device so that a continuous loop back to the sequencer could be made. In this day and age, multiples of MIDI INs and OUTs are possible using modern MIDI interfaces as discussed earlier; therefore, the MIDI THRU socket is less used.

3.12 Avid Hardware Integration with Logic

Avid, the manufacturer of ProTools, produces a number of audio interfaces that roughly fall into two categories. The first of these are what used to be called their LE range, which is a host-based system, meaning that all the digital signal processing is handled within the computer's processor. These plug-ins are known as a Real Time Audio Suite (RTAS). The second category is intended for professional applications. Avid, then Digidesign, pioneered the TDM system that makes use of the DSP power held on their external HD cards, leaving the host CPU to get on with the work at hand.

Both of these methods can be used with Logic as an audio interface using Avid's CoreAudio Manager. The Digi CoreAudio Manager effectively bridges between the hardware and OS X's Core Audio Services, allowing you to control the hardware as though you were in ProTools Hardware Setup page. When using the Avid CoreAudio Manager with Logic, it is worth noting that Logic and the Avid interface may not immediately talk to each other. Opening up Avid's CoreAudio Manager before launching Logic should ensure communication.

It should be noted that this connection is not the best using Logic for recording as typically there can be significant latency delays. For mixing and listening back, the Core Audio solution is perfectly workable.

KNOWLEDGEBASE 2
ReWire

ReWire is a system developed by Propellerhead, the makers of Reason and ReCycle. ReWire cleverly allows internal connections to occur between audio software packages within one computer. It is therefore possible to have multiple MIDI connections internally between applications and receive the corresponding audio signals back in return, again internally. This is an incredibly powerful and flexible solution when getting the best from two applications. Setting up ReWire in Logic is very easy. You must ensure Logic is opened before the ReWired application is started.

With Logic opened, launch the ReWire application, thus ensuring that Logic knows that this is to be the ReWire application and will therefore make the appropriate connections for you. To use your ReWire instruments, first begin to create an External MIDI track by pressing Cmd + Alt + N on the Arrange area. Next, call up the Library Tab from the Media Icon on the top right-hand side of Logic. Within the Library should be your opened ReWire application – in this instance, let's say Reason. Clicking on the Reason folder in Logic's library should reveal all the open Reason instruments. Simply click on one to map this to the selected MIDI track.

As you select new instruments within your ReWire application, both the MIDI and audio connections should be automatically made to Logic. The audio connections will appear as Auxiliary tracks on your Logic Mixer and can be processed using Logic's plug-ins and blended into the mix as though they were another virtual instrument.

In this chapter

Knowledgebases

Plug-in Focus

Starting a Project

4.1 Introduction

Laying the successful foundations at the start of your project is vital to the long-term success and seamless workflow of realizing audio and music production work in Logic. Of course, not all projects start the same way, but we're going to begin by looking at the process of recording audio into Logic – in effect, using it as a modern day replacement to the multitrack tape machine. This is also an excellent way of covering some important essentials of working in Logic: like creating tracks, using the transport controls, the bar ruler, and the basic monitoring. We'll also start to explore the various aspects of data management involved in working with Logic – more specifically, how and where you should store your project data, and what a Logic project is actually composed of.

If you're short of inspiration or just needing something to kick-start your creative process, we'll also take a look at importing and working with Apple Loops. Apple Loops – contained within the multitude of Jam Packs installed with Logic Pro X – are an instant resource of loops, instrument, and vocal passages, all copyright-free and ready to be used in your compositions.

Besides the basics of track laying, we'll also review various ways with which you can make a recording session run even more smoothly – either providing a dedicated headphone mix for musicians, for example, or using plug-in effects to better understand how your instrument might eventually appear in the mix. All these factors will create a more comfortable recording experience for you and your musicians, allowing the musicianship, engineering, and performance qualities to become more focused.

Of course, this is only the starting point to your creative work in Logic, as we'll also be looking at how to edit this newly recorded material in Chapter 5 and then look towards integrating the range of virtual instruments and MIDI sources in Chapter 6.

4.2 Assets and Projects

The basic currency of Logic is a Project, which will contain all the relevant assets for the song you're creating. In respect to Logic, the term "assets" corresponds to the collection of separate files and data that a song can be composed of. For example, in a song file, which contains all the important arrangement information, you could easily assemble any amount of audio files, sample data (for the EXS24 sampler), and movie files, all of which will reside in the main project folder. By organizing your data in this way, both you and Logic can keep better track of your creative process, making backing up, or moving to a different Logic Pro X setup, for example, considerably easier.

The starting point of your creative process is to create a new project (File > New). Logic will then present you with a series of possible starting points, including some interesting options for composing or producing music in Logic, based on either genre or production objectives, like surround mixing, for example, or mastering. If any of these templates meet your particular needs, then it might be appropriate to select them, otherwise select the Empty Project as a suitable initialized starting point for production. Note that you can also automatically select this option by holding down Alt as you select New from the File menu. Assuming that you've selected an Empty Project as your starting point, Logic will prompt you to create an initial track to work with. To get the ball rolling, select Audio under the track Type option and press the Create button to insert a single track. We'll take a deeper look at creating tracks in Section 4.3.

Figure 4.1 Creating a new project (File > New) will present a number of starting templates. Select Empty Project if you want a simple initialized starting point.

Save As ...

Before you go any further, it's worth saving your project, if only to allot a particular location for your project folder. If Logic hasn't already opened a Save As dialog, you'll need to do so (File > Save As …). Besides specifying the name of your project, you can also instruct Logic to automatically save the relevant assets to your project folder, under the "Copy the following files into your project" option. Obviously you need to trade off excessive duplication of data at this point, alongside a clear data management strategy. If you're unsure, there's no harm in selecting all the options, but this will use a correspondingly larger amount of drive space.

In addition to decisions about assets, you'll have the option to either organize the project in a Folder or as a so-called Package. If you want to avoid confusion and clutter, the Package option is useful, as the data (sound files, movies, EXS instruments, and so on) appear in a single combined file. By picking the Folder option, though, you get to see each file discretely organized into an array of different folders. The Folder option is obviously useful if you want an overview of what's happening in respect to data usage, and potentially, the option of managing some of the data usage via Finder.

Although this initial project management can appear time consuming (especially if you just want to get on with making music!), it's worth pointing out that good data management is a vital part of music making on a computer. Spending time to ensure that you know where your data is being stored will reap plenty of dividends in the long run, making it far easier to recall projects several months down the line, or indeed, work more effectively between different computers.

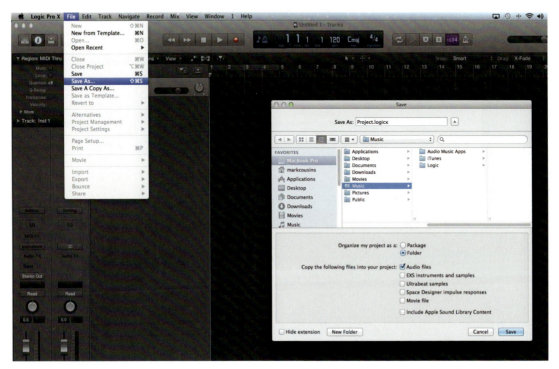

Figure 4.2 The Save As dialog allows you to create a root folder for your project and to check what assets will be saved with it.

Alternatives

The final layer of data management to consider is that of Alternatives. With the development of most songs, it's often the case that you'll want to pursue a different creative goal at some point in the production process: maybe a different mix to please the record company, for example, or a short, 30-second version of your track for use on a commercial. Although the end output is different, the same pool of assets are common to both examples.

The idea with Alternatives, therefore, is that you can explore a completely different version of project without having to save off the project under a new name. You can find the Alternatives option under the file name, where you can see both the current version of the project you're working with alongside other alternative that you might have created. To create a new alternative, use the menu option File > Alternatives > New Alternative, and you can also use the menu option File > Alternatives > Edit Alternative option to name or delete existing alternatives.

Figure 4.3
Alternatives
reference the same
assets, but allow you
to take a completely
different approach
with your project.

KNOWLEDGEBASE 1

What Are Assets?

Assets are Logic's collective term for the various data files that can be associated within a project. Of course, the project file – or song – is only the beginning of the data you'll amass over time, with the full set of assets including audio files, movie files, EXS24 sampler instruments, ultrabeat samples, and space designer's impulse responses. Ultimately, if you don't have some organized strategy for looking after these files, you'll soon find that a project becomes difficult to manage, especially when it comes to either archiving a project or moving it to another system.

By keeping the files in a single folder, Logic offers at least one way of organizing these data, although if you've got an alternative system, you can always decide to manage this yourself. Theoretically, this can prevent certain unnecessary file duplication, which can easily become an issue with large EXS24-based projects. To change the current project's Asset settings at any time, select File > Project Settings > Assets, where you can, accordingly, switch in and out the different parts of the asset management. By default, Logic concentrates on audio files, including automatic sample rate conversion of files imported that don't match the current project settings.

(Continued)

Figure 4.4
Managing the assets related to your project will allow you to create better project archives and to move between different Logic setups in a more effective way.

4.3 Working with Tracks

Assuming you've selected an Empty Project, the next task is to create a series of tracks to record your audio and MIDI information into Logic. For example, if you're recording a band using a number of microphones, you'll need to create an accompanying collection of audio tracks representing each instrument or microphone you're using. If you intend to create a track largely using virtual instruments, with a few vocal overdubs, you'll want to create 16 or so instrument tracks combined with one or two audio tracks. Of course, you can always add, remove, or reorder tracks within your project at any point in the production process, so don't feel obliged to preplan your complete track list at this point.

To create a track, click on the small + sign on the top of the track list, or, if no tracks are present, Logic will ask you what tracks you initially want to create. The dialog box that accompanies track creation allows you to specify both the exact type of track that is being created, and several options for creating multiple tracks, output and input assignments, and so on.

Figure 4.5
Create a number of empty tracks to correspond with the instrumentation you want to record. The New Tracks feature includes several ways of speeding up this process.

Audio

An audio track will allow you to record an audio signal being fed from a corresponding input on your audio interface. This is the primary focus in this chapter.

Software Instrument

Use a software instrument track if you want to record music using any of Logic's integrated virtual instruments (Retro Synth, EXS24, Vintage Electric Piano, ES2, and so on), or indeed, any third-party Audio Unit instruments. We'll be taking a closer look at the process of working with virtual instruments in Chapter 6.

Drummer

Drummer is Logic's in-built solution for the creation of realistic drum tracks. Rather than programming a rhythm track from scratch, you can specify a number of performance criteria (including the choice of drummer) to have Logic intelligently create a drum track for you. The Drummer feature works in conjunction with the Drum Kit Designer plug-in, both of which we'll explore in greater detail in Chapter 6.

Guitar or Bass

The Guitar or Bass option is much the same as the audio option, although in this case Logic automatically opens the Library and loads a default "guitar" setting with amplifier cabinets and effects pedals. In theory, there's no distinct operational difference to that of an audio track.

External MIDI

External MIDI tracks correspond with MIDI hardware like synthesizers and samplers that you have externally connected to Logic (usually through a USB interface). Again, we're going to spend more time looking at this functionality in Section 6.5.

As you're creating your first set of audio tracks, you'll notice a couple of important options to speed up your workflow. For example, if you intend to record from the first 16 inputs of your audio interface, you might wish to increase the number of tracks parameter to 16 and check on the Ascending option as part of the Input drop-down menu – that way, each subsequent track will take a different input number. On the whole, most users tend to monitor their session from a single output, so you'll probably want to leave the output setting out of its Ascending option. Whether its format is mono or stereo will depend on the type of signal input being fed to your audio interface – a pair of drum kit overheads, for example, will be stereo, whereas a vocal (recorded with a single mic) would be mono.

With your basic tracks created, you'll also want to lodge some initial track names. Not only this will make the navigation of Logic's Tracks area and Mixer much easier, but you'll have better chance of keeping track of the various project audio files with names like "kick drum" rather than "audio 01." To name a track, simply double-click on the existing track name as part of the track list. Once named, any subsequent recordings on that track will adopt the name accordingly.

Figure 4.6
Naming tracks will allow you to keep a better handle on your Logic arrangement and is being used to automatically name the audio file and region that you create.

4.4   An Introduction to the Audio Mixer

Although we don't need to go too deeply into mixing at this point, it is worth introducing the concept of the audio mixer, mainly as an alternative way of looking at the tracks you've created for your project. While the track lanes of the Tracks area work horizontally, the Mixer area shows us our current project in the form of a traditional vertical mixer, in which each track is represented as a different channel strip. You can open the Mixer at any point using the Mixer tab either at the bottom of Logic's main window or from the keyboard shortcut X.

Common to both the Mixer and Tracks area's track list are a series of function switches, each of which govern how the particular track or channel is working at any given point in time.

Figure 4.7 The Mixer window (accessible from the Mixer tab or by pressing X) displays an alternative view of your newly created tracks.

Figure 4.8 Input Monitor

Input Monitor (I)

Input monitoring is a simple one-click means of directly monitoring, or listening to, the current selected input for that given track or channel. At first, this might be a good initial way of establishing what's coming into your inputs, and, using the meters on the channel strip, setting an initial gain on your microphone and line inputs. Note that in most cases, this will be done directly from the interface itself (or the interface's control software, if you're using something like the digitally controlled Apogee Ensemble) rather than adjusting the fader's position. In this case, the fader is simply setting the monitor level, rather than any form of recording level as it might do on a conventional console.

Should you wish to change a track's particular input, this can be done from the Mixer, clicking on the small Input slot halfway down the channel strip. The list presented should tally with the inputs available on your audio interface. For example, a typical USB audio interface will present two available inputs, whereas a FireWire interface could offer ten or more possible inputs. Clicking on the small icon button to the side of the Input slot will also change the input format between mono and stereo operation, with a corresponding change in the input selection – so that input 1, for example, would become input 1–2.

Figure 4.9 Use the I/O box on the channel fader to change a track's input status, with the number of inputs corresponding to the type of interface you're using with Logic.

Record Arm (R)

Record arming a track or channel indicates to Logic that you're about to make a recording. Again, you'll be able to see your input levels accordingly and hear the signal routed through to your main outputs. Once you press record on the transport, Logic will actively record the input as a new audio file (more on this in a moment!).

Mute (M)

As you'd expect, mute silences the track or channel in question. As the controls between the Tracks area and the Mixer are linked, muting a track in the Tracks area will also mute the corresponding mixer channel strip and visa versa.

Solo (S)

Solo will play the chosen track or channel in isolation, temporarily muting all other tracks in the project. Note that you can also add to an existing solo

Figure 4.10
Record Arm

Figure 4.11 Mute

Figure 4.12 Solo

selection so that you hear several tracks in isolation. Alt-clicking on an existing solo selection will remove all the selected tracks from solo mode, returning the mix to normal.

Working between the Mixer and the Tracks area is an important part of the day-to-day usage of Logic. For example, note how a selection in either window is mirrored in the other – selecting the vocal track on the Tracks area, for example, will also bring it up on the Mixer.

4.5   Using the Transport Controls and Ruler

As with any multitrack recorder, the Transport Controls – found as part of the control bar – provide the master control for recording, playing back, and looping in Logic. For example, having created your tracks and record-armed them, pressing Record on the transport will tell Logic to engage its playback (of any existing material) and record each track as an individual audio file. The rest of the controls (play, pause, rewind, fast-forward, and so on) need little introduction, although it's worth noting how the transport controls are also influenced by settings in the Tracks area's bar ruler.

On the right-hand side of the control bar, you'll find a number of different Mode and Function buttons – like a cycle switch, tuner, solo button, and of course, the all-important Metronome. We'll be discussing these in more detail as we move through the recording functions in Logic.

Figure 4.13 The control bar also includes a number of important Mode and Function buttons – including the Metronome and Autopunch – that are vital to the process of recording in Logic.

Ruler

The ruler – along the top of the Tracks area – provides the positional reference, or timeline, for our project, with the Playhead indicating the current song position. As you'd expect, there are a number of additional functions of the ruler, allowing you to specify the parts of the recording to be dropped in, for example, or areas to be looped around. Crucial to the functionality are the left- and right-hand locators, which are indicated by the yellow shading (assuming the cycle mode is active) on the ruler. If cycle isn't active, you'll still see the position of the locators – this time using a subtle difference in shading.

Figure 4.14 Here are two left and right markers defining the length of the cycle. Use the cycle feature either to loop round a portion of the song or to perform a series of different takes.

You can adjust the left and right locators in a number of ways, although probably the quickest and most logical solution is to drag from left to right (with the mouse button held down) in the top half of the timeline. As you drag your selection, you'll automatically place Logic into cycle mode – as illustrated by the yellow shading in the ruler, and the fact that the cycle mode indicator (found in the right-hand side on the control bar) is active. In cycle mode, you can repeat over a section of the music, and if you're recording, lay down a number of different passes or takes of the same musical line to be edited later on.

To disable cycle mode, you can either click on the yellow region in the ruler or alternatively deactivate cycle mode on the control bar. Even when it's deactivated, Logic will retain the left- and right-hand locator settings allowing you to return to the cycle at any point. Furthermore, you could also define a number of different cycles using the Marker feature (like Verse, Chorus, and Middle Eight), which we'll discuss in more detail in Chapter 10.

4.6   Your First Recording

Assuming you've created and record-enabled the appropriate amount of audio tracks, you should be ready to make your first recording. Recording will begin in one of two places – either at the current position of the playhead or, if you've set up cycle playback, at the point the cycle begins. To provide you with a small amount of preroll, Logic will actually start playback a bar earlier than the current record point (assuming you have Count-In button engaged), allowing you to get some indication of the track's speed, and existing elements of the performance that might lead up to it.

When you've finished your first pass, take the track out of record arm mode. If you've got multiple tracks record-enabled, this can be done by Alt-clicking on any one of the active record-enabled tracks. Playing the project back will now allow you to audition the takes, spotting any deficiencies or areas you might want to improve.

Figure 4.15 Recording the first pass into Logic.

KNOWLEDGEBASE 2
More about the Metronome

In certain situations, you might find it beneficial to adapt some of the qualities of the Metronome – stopping the Metronome being sent over MIDI, for example, or changing the preset Klopfgeist sound to something of your own liking. The Metronome settings can be accessed either through the project's settings (File > Project Settings > Metronome) or directly by Ctrl-clicking on the Metronome icon on the control bar. The settings allow you to switch the Metronome in and out of MIDI playback, as well as being able to disable the software click and adjust some basic parameters like tonality and volume.

By delving deeper in to the audio mixer environment (which we'll discuss in more detail in Chapter 11, but for now access through Window > Environment), you can also change the default Klopfgeist instrument assigned to the Metronome. Klopfgeist is a simple instrument, specifically designed to produce distinctive click sounds that can be audible over a mix, yet not so loud that they would bleed out excessively over headphones. If you've got the Mixer environment open, you should see the Klopfgeist instrument assigned to the Click object. By changing this to the EXS24 sampler, for example, you could replace the standard Klopfgeist click with a rimshot sample from one of the factory kits.

Figure 4.16 Adapting the Metronome settings can be achieved through the menu option File > Project Settings > Metronome.

4.7   Overdubs and Punching-In and -Out

Although we'd all like to think we could record perfect "one-take wonders" every time we enter the studio, there's always an inevitable amount of remedial recording that needs to take place – either with fluffed notes or whole sections that need to be replaced. Logic, of course, provides a number of solutions for this, all of which can be further refined when it comes to editing your project (as we'll see in Chapter 5). Unlike tape, it's worth remembering that any punch-ins or overdubs are "nondestructive" – in other words, the original takes remain untouched on the hard disk, with Logic simply changing or augmenting the files that it decides to playback for any given track.

Quickly Punching-In Material

One solution to repairing your recording could be simply to record over a section of the track in question. To do this, all you'll need to do is place the track or tracks you want to replace into record-ready mode, locate the playhead at the punch-in point, and press record. Logic will then give you the usual bar preroll and then initiate recording, effectively "replacing" any previous recording you have made at that point

Figure 4.17 If you want to replace any section of the recording, simply locate the transport at the point where you want to drop in, and press record. Logic will provide a bar of preroll and then begin recording your new material.

(although remember, this is nondestructive). When you press stop, you'll effectively punch-out, leaving subsequent audio recordings untouched.

Recording in Cycle Mode

If you've got a specific section of the song you want to work on, it might be appropriate to record in cycle mode. In cycle mode, Logic will continually loop the portion of the song, recording each pass that you make. When you've stopped the transport, the last take will be used for playback, although all the various takes will have been recorded.

Figure 4.18 Using the cycle function is a useful way of loop-recording a number of bars until the part is performed correctly.

Autopunch

If you're clear about a specific part of the track you want to replace, consider using the Autopunch feature. Autopunch can be activated through the control bar, with an additional set of locators (shaded red) viewable in the middle of the bar ruler. By default, Autopunch isn't initially visible on the control

Figure 4.19 If you have a specific part of the take you want to replace, consider using the Autopunch feature for an accurate, automatic drop-in and drop-out.

bar, so you'll need to Ctrl-click on the control bar, select Customize Control Bar and Display and enable Autopunch for the selection of objects. You can change the position and length of the Autopunch locators in the same way as the cycle locators, ideally setting the locator points to precisely match the portion of the track you want to replace. With the track record-enabled, and record engaged on the transport, Logic will playback the song and only record for the precise duration of the Autopunch locators.

Introducing Take Folders

Whenever you overdub onto an existing recording on an audio track, or use the Autopunch feature, you create what Logic refers to as a take folder. If you look closely at a take folder, it looks slightly different to a conventional audio recording in Logic, with two triangles in the top left- and right-hand corners of the region. Technically, a take folder stores all your various takes in an organized way, with features that allow you to switch between takes or to make quick, on-the-fly edits between the different passes.

To understand how we can carry out further work with the take folders, take a look at Section 5.4.

Figure 4.20 If you record over an existing region, Logic will create what is known as a Take Folder. A Take Folder will allow you to reposition and refine edit points later on in the production process.

4.8 Creating Further Tracks and Track Sorting

Although we should have created the majority of tracks required for our session at the start of the project, it is highly likely that you'll need to add further tracks into your session. Let's take a look at some of the motivating factors and the quick steps to create and organize these tracks.

Creating New Tracks for the Same Instrument

Use the menu function Track > Other > New Track with Same Channel (or Shift + Ctrl + Return) to create a parallel track with the same instrument or channel assignment. This is a useful tool if you want to create two alternate "takes" of the track – so, rather than packing different takes in a Take Folder, you end up with two discrete takes assigned to the same channel. Remember that only one audio recording can take priority at any given point in time, so you'll need to be clear about what track you're listening to by muting the unwanted takes accordingly.

Creating Tracks with the Next Instrument

Creating tracks with the next instrument or channel strip (Track > Other > New with Next Channel) is a great way of augmenting your existing project, adding further track lanes for each new instrument you might want to add into the mix. Using the keyboard shortcut, Ctrl + Return, allows you to do this quickly and efficiently without interrupting your creative workflow.

Figure 4.21
Creating a track with the same channel is a good way of negating the Take Folder system, allowing you to clearly visualize each take as it's being recorded.

Creating Tracks with Duplicate Settings

If you've already set up some particular monitoring EQ and compression effects, for example, you might want to create new tracks with duplicate plug-in settings (Track > Other > New with Duplicate Setting, or Cmd + D). The newly created track will have its own unique channel assignment, but a copy of any of the plug-in settings as well as any fader or pan positions you've established on the original track.

Sorting and Organizing Tracks

As the project grows, it might become beneficial to organize your track list into an order that better corresponds to the arrangement of instrumentation, as opposed to the order you recorded it in. For example, maybe you want to sort all the vocal tracks together with the lead vocal first, or separate the keyboard from the guitars. Arranging the track order is as simple as clicking on a track lane and shuffling it up or down the track list accordingly. Should the track list become increasingly confusing, you might also want to return it to something closer to its original arrangement. In that case, go to Track > Sort Tracks by > Audio Channel, for example, and have the tracks automatically reordered by their channel assignment.

Figure 4.22 Try sorting tracks to make their organization clearer.

4.9　The Project Audio Browser and Importing

Although the Tracks area provides an important positional overview of the audio files we recorded as part of our project, it's not the only way we can manage and organize the various files that we've created. The Project Audio Browser presents the complete list of the audio files included in your session and offers a means of importing new audio files (possibly from another project, or a drum loop from an audio CD). To open the Project Audio Browser, select View > Show Browsers, or open the Project Audio Browser using the keyboard shortcut F.

Scrolling down the Project Audio Browser, you should be able to see all the files that are contained within your project, even the ones that might not be currently active in your Tracks area. Alongside the file name, you can also get some basic information on the audio file, including its sample rate, bit depth, and file size. Another important piece of information is the file's location, which is displayed in the top line whenever a particular file is selected. If, as part of your asset management, you specified for all files (recorded or imported) to be stored in your project folder, then all these locations should be the same. If the project's data management isn't quite so organized, this might be an opportunity to find out which files are located outside your main Audio Files folder.

Figure 4.23 The Project Audio Browser provides an effective overview of all the audio files used within a project.

Importing Other Audio Files and CDs

To import existing audio files, you can also use the Project Audio Browser. Logic supports a number of different audio file formats – including WAV, AIFF, SDII, CAF, MP3, AAC, SMF, and REX2 – all of which can happily reside together in the same project. To import a file into your project, go to Audio File > Add Audio File and browse your drives accordingly. Again, if you've got your assets setting established correctly, the files will be copied into your project's Audio Files folder to keep all the media content in a centralized, project-specific location.

To finish the importing process, you will, of course, need to drag the audio file into your project's arrangement. The precise method you choose to do this will have an effect on how the resultant file is played back. If you drag the file onto an existing audio track, the new audio file will be played back with the settings on that current channel strip. However, if you drag the new region to a position in the Tracks area without a track assignment (this would be at the bottom of the track list), Logic will create a new track specific to the audio file in question.

Figure 4.24 Use the Project Audio Browser to import additional audio files into your project. These can then be dragged into your arrangement with a new track created accordingly.

4.10   Working with Apple Loops

Apple Loops provides a quick-and-easy source of inspiration for music and audio production, and can be an effective way of kick-starting a project. Included with the full install of Logic Studio are a number of Jam Packs providing everything from urban drum loops to Foley effects. The important point to note about the Apple Loops format is that the tempo and/or the key of the loop will automatically conform to your project's current settings, allowing you to quickly audition a loop's relative merits without having to time-stretch the file to fit. Equally, if the project's tempo changes at a later point (even midway through the song!), the Apple Loop will follow these changes accordingly.

You can access the Apple Loop browser via the control bar, or use the menu option View > Show Apple Loops (keyboard shortcut O). The Apple Loops browser is an extremely effective way of exploring the content installed with Logic Pro X, allowing you to specify various search tags to sift through the huge amount of possible sound content. The browser itself can be organized into two different viewing styles – a Button view or a Column view. Arguably, the most intuitive of these is the Button view (indicated by the small musical notes), which organizes the various loops by genre (Jazz and Country, for example), instrument (Vibes, Strings, and so on), and sound-type "categorization" (Melodic, Distorted, Clean, and so on). By clicking on combinations of these tabs you should be able to arrive at a loop or collection of loops that best suits your needs. Pressing the Reset tab (top left) will remove any existing selections, allowing you to perform a new search.

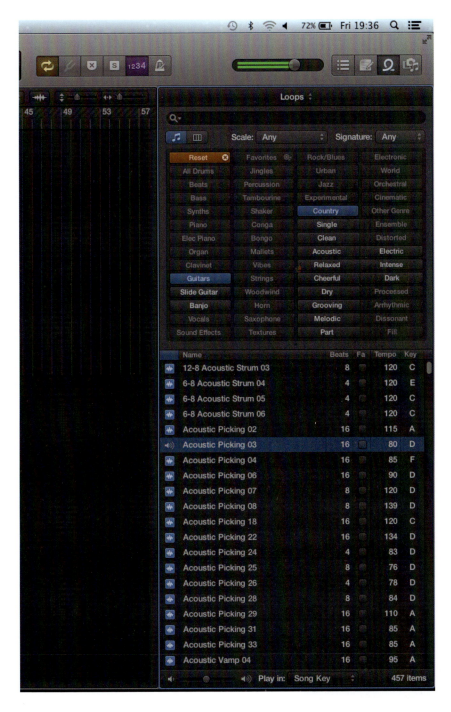

Figure 4.25 The Apple Loop browser allows you to quickly locate different loops based on their musical qualities.

The alternative browsing method is the Column view, which sifts through the sound content in a way similar to the Finder in its column mode. As you'd expect, both modes can also use a search string if you're looking for something specific in mind.

Figure 4.26
The Column view provides an alternative way of navigating and viewing your Apple Loops sound content.

To hear an Apple Loop, simply click on its file name, with Logic playing the file back at the same tempo and key as the song you're working with. You can also use the small volume slider and the bottom of the loops browser to set the volume for auditioning the loop. It's also worth checking the percentage match, as this will have some effect on the relative quality of the loop in the final arrangement, with loops closer to the project's tempo requiring less time compression or expansion to get them positioned correctly.

KNOWLEDGEBASE 3
Importing REX2 Files

The REX file format was originally developed by Propellerhead as a cross-platform solution for transporting time-sliced audio files created by their popular ReCycle application. A number of third-party loop libraries make use of this file format, so it's useful to know how Logic handles these files.

One option is to load the REX file directly into the Tracks area itself, with the slices left as a series of different regions locked to your particular project tempo. To import the loop, go to the Browser and drag the REX file into its first position in the Arrange area. Logic will then ask you to confirm the REX file import, possibly suggesting a series of crossfades to best match the original loop tempo with the tempo of your session. If you confirm, this will create a folder in the arrangement containing all the accompanying REX slices.

As an alternative to importing the slices, you can also choose to render the REX file as a complete file or an Apple Loop by selecting the appropriate option from the pull-down menu. Arguably, a rendered audio file is easier to manage than a collection of slices, although it won't be able to follow changes in the project's tempo. As an Apple Loop, though, the REX file will retain its elasticity, allowing it to match the project's tempo at any point in the creative process.

Figure 4.27 REX files can be directly imported into your Logic arrangement, with a number of different options to conform the loops to your project's tempo setting.

Audio Files versus MIDI Files and Instruments

The Apple Loop files format actually contains two variations – a pure audio file that offers a recording of the loop in question and a MIDI version (also known as an Apple Loop +) that combines the raw note data with an accompanying software instrument and effects. You can see what type your Apple Loop is by looking at the small icon to the left-hand side of the file. A blue waveform icon indicates that the Apple Loop is only available as an audio file, whereas a green music note indicates the Loop is available as raw musical data (in other words, a MIDI version) and as an audio file.

In the long run, a MIDI-based version will provide much greater flexibility than an audio-based Apple Loop, allowing you to precisely edit the musical data, instrument settings, and effects. However, these versions can, at times, make slightly larger demands on your CPU usage.

Figure 4.28 Here are the two Apple Loop file formats and how they differ when imported into your session. A standard Apple Loop (top) is audio-only, but the enhanced version (bottom) contains the specific MIDI and instrument data used to create the loop.

Importing Apple Loops into Your Session

To use either of the two Apple Loop formats, all you need to do is simply drag the file into your arrangement. One point to be clear on, though, is the type of track you import it to, and how this has an effect on the end form of Apple Loop imported. For example, if you drag the Apple Loop onto an empty track lane with no track assignment, Logic will automatically create the appropriate type of track (Audio or Instrument) based on the particular Apple Loop in question. By dragging the Loop to a specific type of track, though, the Apple Loop will be conformed to the track type – so, for example, MIDI-based Loop would be placed as its audio version on a corresponding audio track. The only exception, of course, is that audio-only loops can't be converted into MIDI regions.

4.11   Improving What the Artist Hears – Headphone Mixes

Although we've looked the process of basic track laying in Logic from the engineer's perspective, it's worth considering what exactly the artist hears and performs to as the material is being recorded – in respect to either the individual headphone mix or the additioning of plug-in effects (like compression, distortion, or reverb) – that might be added during the process of recording. However, it should be mentioned that these subsequent techniques will push your Logic skills much harder than that with respect to the use of the Mixer that has been discussed in the first part of this chapter.

First off, let's consider the two types of mixes heard in a typical tracking session in a professional studio.

Monitor Mix

The monitor mix is simply the mix that is heard in the control room and is usually based on a reasonable and considered balance of the material that has been recorded. In some cases, the producer or engineer might want to hone in on a specific part (usually whatever is being recorded at that point) with the rest of the instrumentation significantly quieter. Ultimately, a good monitor mix is the only way you can ensure that you're making an effective recording.

Cue Mix

The cue mix (also known as talent mix or headphone mix) is the balance of material that relates specifically to the performer and what they hear through their headphones.

Of course, a pragmatic solution to recording sometimes involves the monitor mix and the cue mix effectively being the same thing, especially if you're recording yourself! If you're using a USB audio interface, it might also be the case that the interface features some direct provision to control the mix between the input signals and Logic's mix so as to combat any latency issues (for more information on latency see Chapter 3), again resulting in a combination of both the monitor and cue mixes. However, if your audio hardware supports it and you need a more refined monitoring system, it's far better to create a separate monitor mix and cue mix directly from Logic.

Creating a unique headphone mix in Logic involves a more detailed use and exploration of the Mixer area, so as to effectively create two mixes at the same time. On each of the channel faders in your project, you should find an empty box next to the Sends section. As we'll see later on the mixing chapter, these bus sends, as they are called, have several different applications, but for now, we're going to concentrate on their role in the creation of a cue mix. To create a bus send for a particular channel strip, click in the Sends slot and insert a send to bus 1. Note that if you have a large number of tracks, you can also select multiple channels, insatiate the bus on one of the channels, and have them all duplicate the same action.

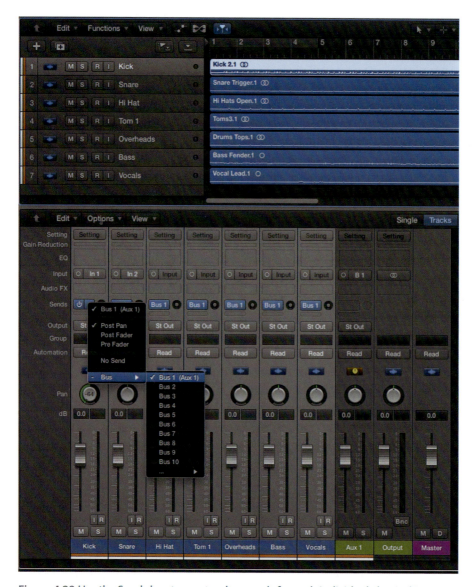

Figure 4.29 Use the Sends box to create a bus sends for each individual channel in your mix. The bus sends are the controls that will be used to create your headphone mix.

The next important thing to do is to set the pre/postassignment of the bus sends. Although this sounds complicated, all a post/preassignment means is whether the bus send is patched before or after the channel fader. Thinking through logically, it makes most sense to create the cue mix pre fade – so that any movement of the channel faders won't immediately change the relative mix for the headphones. In other words, if you want to turn down the vocal in your monitor mix, the vocalist can still hear himself or herself loudly and clearly. The pre/post assignment can be easily switched by clicking on the bus send and changing the mode from post (the default setting) to pre. If this is done correctly, the position of the send should switch to the left of the send slot.

In creating the bus send, Logic will have also created a corresponding aux master object, which will act as the master fader for you cue mix. To make things clearer, you could double-click on the channel's strip name toward the bottom on the Mixer and change this to something like *headphones* or *cue mix*. To build up the headphone mix, solo off this channel and then blend the appropriate instrument using the bus sends on the individual channels. As a good tip, maybe start by placing the loudest possible instrument in at around 0 dB (you can do this by Alt-clicking the corresponding bus send pot) and then positioning the rest of the instrumentation around this.

To send the headphone mix to the specific physical interface output, change the output status (found just below the Sends section) of the aux master accordingly.

Figure 4.30 Switching to prefade will allow the mix to be completely independent of Logic's channel faders. The color of the bus sends indicates the pre/post status, with green being prefade.

4.12   Monitoring through Effects

In some cases, performing a take can sometimes be tricky without an idea of how it will sound in the final mix. Although it is often the intention of the engineer to record "dry" sounds devoid of artifacts like compression or reverb (which will instead be added in a controlled way in the mix), the same cannot be said of what the artists themselves prefer to hear. Far from being fussy, effects like reverb can actively help a vocalist pitch themselves, equally the drive and distortion of a guitar sound is such an integral part of how a guitar is "played."

Effects plug-ins can be instantiated using the small gray Inserts boxes in the Mixer's various channel strips. Clicking on the box will bring up a list of Logic's own internal plug-ins (organized into groups – like delay, distortion, filter, and so on), as well as any third-party Audio Unit plug-ins that you have installed on your system (organized by manufacturer). You can also activate EQ at any point by double-clicking on the corresponding EQ box just above the inserts. An important point to note is the order of the plug-ins – reflected in the progression from top to bottom on the inserts – which will dictate the order in which an instrument is processed. If you're a guitarist, for example, you might want to place certain effects before and/or after your virtual amplifier (in this case, Amp Designer). Alternatively, a vocal track tends to have EQ and compression before reverb, so that the reverb isn't equalized and compressed.

Figure 4.31 Recording through an effect plug-in will have no direct impact on what is being recorded, only what you and the musician hear in their mix.

We'll be taking a closer look at the art of setting up plug-in effects on inserts, and how their order affects the processing taking place, in more detail in Chapter 8.

The point to remember with plug-ins instantiated in this way is that they have no direct impact on the actual signal that is recorded to disk. Although this might first seem a little frustrating, it means that, if you change your mind later on, you can always return back to its initialized state – peel back the layers of distortion, on a guitar, for example, and you'll still have the completely clean, DI'd version. In effect, these plug-ins are used exclusively for monitoring purposes – they're keeping the musician happy with what they hear in the monitor mix, and the engineer content with what is being printed to disk.

Accounting for Processing Latency

Although most of Logic's own plug-in are optimized for low-latency operation, it is possible that creating long chains of plug-ins, as well as using certain third-party plug-ins, might start to add unacceptable amounts of latency for real-time recording and monitoring. If playing starts to become unresponsive, you might want to consider removing certain plug-ins, although these can always be reinstated once it comes to the mix (when latency isn't an issue).

As an alternative to manually switching plug-ins in and out, try using Logic's own low-latency mode, which is available directly from the control bar (although the button isn't viewable by default, so you'll need to customize the control bar first), or under the General Tab of the Audio Preferences (Logic Pro > Preferences > Audio), where you can also modify the accepted latency limit from the default of 5 ms. Using the low-latency mode is an excellent way of being able to quickly move between tracking and mixing tasks, without having to worry about which plug-ins are causing latency problems.

Figure 4.32 Using the low-latency mode will ensure that any DSP-hungry effects plug-ins won't create unacceptable amounts of monitoring delay.

PLUG-IN FOCUS

Amp Designer

Amp Designer is a virtual presentation of a guitar amplifier, including the amplifier head, speaker cabinet, and virtual microphone used to "record" the amplifier's sound. As such, Amp Designer is an essential plug-in if you're recording with an electric guitar plugged directly into the instrument-level input on your audio interface (also known as the DI input). The plug-in transforms the otherwise sterile tone of a DI'd electric guitar into something far more playable – whether it's a heavily rectified metal guitar, or a clean Fender Twin tone. Although you "play through" the plug-in, the guitar is recorded in its raw state, effectively allowing you to modify the tone of the virtual amp at any point in the production process.

Although it's possible to browse Amp Designer's various presets, you can often have just as much fun swapping-out the various components (using the drop-down menu at the bottom of the interface) to create your own customized sound. An Amp Designer "Model," as such, is a combination of amplifier head, cabinet, and the most appropriate microphone. Moving between the different models, therefore, is a good way of auditioning the principle types of tone available to you.

Of course, a more detailed approach allows you to mix and match the various components of Amp Designer – maybe matching a particularly amp head with a cabinet of your choice. Note that each Amp will distort in a slightly different way – moving between the soft bluesy crunch of a Small Tweed Amp, for example, through to the rich and heavy saturation of the high octane head. The different cabinets "color" the distortion in a variety of ways, using different phase cancellations and high-frequency roll-off between the different designs.

Changing the choice and position of the microphone works in much the same way as in the real world – with a condenser microphone offering the widest and most neutral tone, while the ribbon and dynamic mics sound slightly more colored (although this is no bad thing!). Changing the position of the microphone changes the tone, with a bias toward a bassier tone as you move toward the edge of the virtual speaker.

As you'd expect, Amp Designer makes a great pairing with Pedalboard, which we'll cover in more detail in Chapter 8.

Figure 4.33
By mixing and matching the various components in Amp Designer, you can create your own customized guitar sound.

In this chapter

Audio Regions and Editing

5.1   Introduction

Editing and arranging audio regions has become an integral part of the recording process – whether it's simply tidying up a few misplaced notes or creating the complex "glitch and stutter" edits that are common to many modern production styles. Therefore, in keeping with this increased importance of audio editing, Logic includes a wealth of different editing features – from Quick Swipe Comping to Flex Time editing – all of which unite to offer an almost unprecedented amount of control (and creative input!) in how your eventual production sounds to the listener.

In this chapter, therefore, we're going to take a detailed look at all the editing features that Logic has to offer. Starting off from the basics, we'll first explore how to use simple tools within the Tracks area to start forming the structure of our project – bringing parts in at different locations in the song, for example, or looping two to four bars of music to form a repetitive feature in the track. Exploring deeper, we'll look at the various tools that can be used to improve an original performance – from the ability to "comp" a performance from a number of different takes through to ways of tightening a performance and making it lock more tightly to a predetermined groove.

Of course, there's also plenty of creative potential to be explored using audio in Logic, discovering the intricacies of features like the Audio File Editor, for example, or manipulating audio regions far beyond their original form. As you'll soon see, Logic offers a vast array of creative possibility in audio editing, and effective use of its editing features forms the cornerstone of contemporary music making.

5.2   The Big Picture – Forming Your Basic Arrangement

In Chapter 4, we explored the basic principles of getting audio regions into your Logic project – as audio recordings, prerecorded audio files, Apple Loops, or REX2 files. However, having acquired these assets, we now need to start thinking about how we can organize and edit this raw material together to form our finished song. As you'd expect, the precise way you execute this will vary given the material you're presented with, so you'll need to be conversant with the various approaches and solutions for handling and editing audio regions in Logic.

Broadly speaking, your goals will be focused on two levels of interaction. First, you'll need to pay attention to the overarching structure and arrangement of the project, using the various editing tools to bring instruments in and out of the song. Second, you'll need to investigate the detailed qualities within each performance – like the choice of different takes at various points of the song, for example, through to the timing of each note that is played.

The first simple technique, therefore, is the ability to start moving regions within your arrangement – making copies of regions, looping regions, or muting them. You can move a region simply by selecting it

Figure 5.1 Forming the basic arrangement – either by dragging regions to new locations and copying them (by holding down the Alt key) or muting them.

and moving it to a different position in the Tracks area, either at a new point in the timeline or even onto a completely different track. Holding down Alt key as you perform the drag allows you to create a copy of the region – maybe copying over a chorus vocal line, for example, or a repeated guitar motif, so that it appears at several different points in the track.

As you'd expect, the Tool menu (which can be quickly accessed by pressing the keyboard shortcut T) plays a vital part in the editing process. The application of most tools should be apparent, although at this stage, you'll probably want to principally work with the Pointer Tool (to move and copy regions), the Eraser Tool (for deleting regions), or the Mute Tool (to temporarily disable a region from playing back). We'll be looking at the Scissors, Crossfade, and Marquee Tools later on in the chapter, although it's worth noting their presence for now.

Figure 5.2 The edit tools play an important part in the editing and arrangement process.

5.3   Using the Inspector

In conjunction with the graphical changes made in the Tracks area, it's also worth noting another important feature in Logic's approach to audio editing – the Inspector. As we first saw in Chapter 4, you can find the Inspector toward the left-hand side of the Tracks area, assuming that it has been toggled into view by selecting View > Show Inspector or by using the keyboard shortcut I.

The Inspector's Region Parameters display (seen at the top of the Inspector) shows the current text-based information that accompanies any selected region. The exact information displayed in the Inspector's Region Parameters display will vary given the type of region selected, although given that we're principally working with audio regions at this stage, you should see options that relate to the Delay, Mute, Loop and Gain settings.

Figure 5.3 The Inspector's Region Parameters display highlights the current settings for the selected regions.

The purpose of the Inspector's Region Parameters box in editing is twofold. First, the Region Parameters box provides an extremely accurate way of interacting with a region's current parameter settings – like increasing a region's gain by +1 dB, for example, or applying a fade that lasts just a few milliseconds.

Second, and arguably just as importantly, the Region Parameters box also proves itself useful as a way of applying the same edit across multiple regions – like fading-in all the regions at the start of a song, for example, or time-adjusting multiple percussion parts.

Rather than covering all the applications of the Inspector here, we'll be dealing with various uses throughout the rest of this chapter.

5.4   Working with Loops, Copies, and Aliases

Many compositions feature a large number of repeated regions – like a two-bar drum loop, for example, or a repeated bassline – that occur throughout the duration of a song. In Logic, there're a number of different ways you can deal with repeated musical phrases – from the ability to create multiple "hard" copies to the option of creating instantaneous region loops that last the full duration of the track.

Making Multiple Copies

As an extension of the existing region duplication feature (using the Alt key, in other words), consider using the menu option Region > Repeat Regions (Cmd + R). This allows you to create multiple copies of

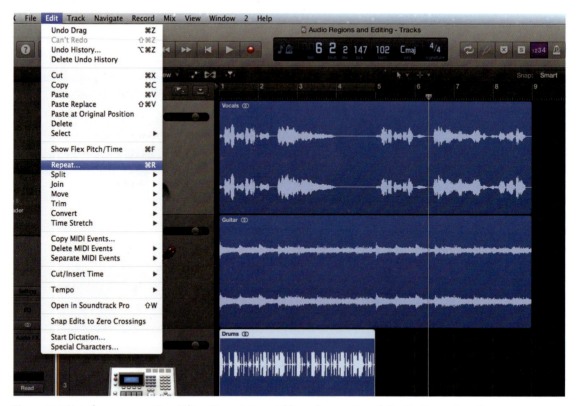

Figure 5.4 Use the menu option Edit > Repeat to quickly make a series of duplicate copies of a selected region.

the selected region/s along the duration of the timeline, specifying the number of duplicates you want to create.

One point worth noting, though, is the ability to distinguish between the "hard" Copies option and an Alias or Clone. In most cases, you'll want to select the Copies options, so each newly created region is a discrete copy in its own right – separate from the original region. With an Alias or Clone, though, Logic will always refer to the original region – so, if you make any subsequent physical edits to the region (like trimming off the last bar, for example), all the other cloned regions will follow this edit.

Loops

As an alternative to using Copies and/or Aliases, consider using the region looping feature. You can apply a loop in a number of different ways, including the Inspector's Loop parameter, the keyboard shortcut L, or by using the unique Loop Tool that should appear at the top right-hand edge of the region. A loop is indicated by a grayed-out version of the region extending to the right-hand side of the original part. By default, a loop will last for the full duration of the song, or until it hits another region on the timeline.

Looping is a great way of quickly extending a song and sketching out an arrangement without having to create multiple duplicate regions. As you'd expect, edits that you make to the original region will be reflected in the subsequent loops. This is much easier to keep track of than the region alias feature (thanks to the use of the shading) and can even be put to use on some interesting creative treatments. One of the best examples of this is creating loops within your song that don't fall on exact bars – like a three-beat phrase, for example, or a five-beat motif. The result creates constant movement in the song, especially when played against a rigid 4/4 loop.

Figure 5.5 Region looping – which is accessible from the Inspector – is a quick-and-easy way of repeating a region throughout the duration of the song (or until the region hits another object).

5.5   Precise Region Editing

Having established the broad structure of your song, it's now time to start to look at some of its finer details – from the precise placement of individual notes, for example, through to the task of "comping" between a number of different performances. Not surprisingly, this will start to push our use and understanding of Logic's editing features much further, utilizing a greater range of editing tools (like Flex Time or Quick Swipe Comping), as well as our ability to handle audio regions at much finer division than just bars and beats. The result, however, will be a significant improvement in the tightness of our performance, and the musicality of the track in general.

5.6   Snapping – Understanding Your Editing Grid

So far, the edits that we've made have been largely based on bar divisions – for example, duplicating two bars of drums throughout the duration of a song, or offsetting a bass part so that it starts on bar five. More refined editing through requires a finer grid to work from – moving from divisions based on bars and beats, right down to individual ticks, or even sample frames! By editing at such a detailed resolution, your edits will be that much more precise, allowing you to accurately position an edit to the most detailed timing resolution the human ear can perceive.

In Logic, the current editing resolution is defined by the Snap setting and can be found toward the top right-hand side of the Tracks area. Options include various different timing divisions (like Bars, Ticks, Frames, and Samples) and an intriguing "Smart" mode that intelligently adjusts the editing grid based on your current zoom resolution.

At the start of a new project, the Snap value defaults to the Smart setting. This is an intuitive way of working with Snap and, to a large extent, provides you with the freedom not to worry about any current

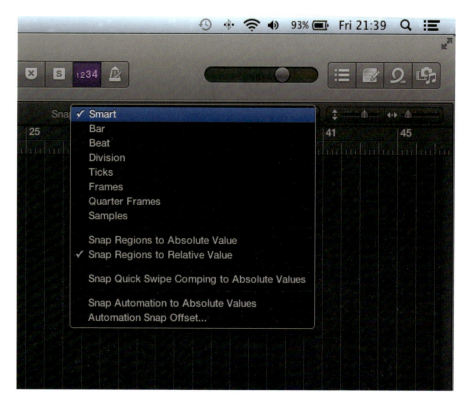

Figure 5.6
Logic features a variety of Snap settings, although the "intelligent" Smart option is best for day-to-day production activities.

Snap settings as you perform your edits. Dependent on the zoom level and the size of the region, the Smart mode interprets the division you require – if you're working at a "macro" level, for example (moving regions around to form the structure of your track), Logic will stick to moving regions in bar increments. Zoom in closer, though (maybe to edit an individual note within a performance), and Logic will work at a correspondingly finer grid – whether it's sub beats, for example, or individual frames.

Another important point to note in editing resolution is that Logic always attempts to keep the relative positions within bars, beats, and so on. Therefore, precise, nonstandard edits can remain relative to the "snapped" grid when moved or copied to other position in the project – a bass part, for example, that picks up 1/16th before the start of bar 17 retains this offset when moved a bar ahead in the track (even if the Snap value is set to Bar). Again, this allows you not to worry about Snap settings and region offsets as you go about making your arrangement, although it's also worth remembering that just because a Snap setting is set to Bar, for example, doesn't always mean that the region will always be precisely placed on a bar.

Figure 5.7 Logic tries to preserve any established "positional offsets" as you move a region around the arrange area.

Going Finer – Bypassing Snap and Nudging Regions

Of course, there are times when you may wish to bypass the snap constraints, even when it's set to the intelligent Snap mode. This can be easily achieved by holding down Ctrl whenever you use a corresponding tool – like slicing a region, for example, or moving a region using the Pointer Tool.

Another option is to use the Nudge option, which is specifically designed so that you can move audio region in small increments. A region can be nudged by selecting it, and then its position can be adjusted using either the left or right cursor keys combined with the Alt key.

As a quick way of adjusting the current Nudge value, try Ctrl-clicking on the region and selecting Move > Set Nudge Value from the drop-down contextual menu. By using this feature, you can quickly set the Nudge value to the most appropriate setting – whether it's as fine as a Sample, for example, or even Bar and Beat settings to maneuver regions about on the timeline using a keyboard rather than a mouse.

Figure 5.8 Using the Nudge feature is a great way of fine-tuning a region's position. Change the current Nudge value by opening up the contextual menu.

5.7 Resizing and Cutting Regions

Having understood how to work with different snap resolutions, let's start to explore some of the finer aspects of trimming, cutting, and resizing regions.

Remember that all the edits that you make on the Tracks area are nondestructive and that once you cut a region, it's always possible to resize the region and restore the original information. This makes it easy to audition possible edits, being confident that it's always possible to return to the original recording at any point in the production process.

Cutting a Region with the Scissors Tool

One of the most immediate ways of editing is to use the Scissors Tool to divide one or more regions into two. An extra dimension to the Scissors Tool is that you can scrub across the timeline to hear where you are about to place an edit. Once you've clicked on the region/s to be edited using the Scissors Tool, simply hold the mouse button down, move the mouse back and forth to "scrub" the edit, and then release the mouse button when you're confident of having found the correct location.

Figure 5.9 Use the Scissors Tool to divide a region in two. Remember, you can scrub the edit point to fine-tune the position of the cut.

Another useful addition to working with the Scissors Tool is the ability to quickly slice a region into a number of equally sized slices – like 1/8th or 1/16th, for example. To cut a region in this way, simply place an edit at the start of the region that equals the same size as the subsequent slices (like 1/16th). However, before you release the mouse button, hold down the Alt key, noting the small "+" icon that should be added to the Scissors Tool. On release, the region should now be sliced into a series of regions all of equal length. Try this technique as means of creating "glitch and stutter" type edits, especially using 1/16th and 1/32nd segments of a performance strategically rearranged and looped accordingly.

Figure 5.10 Holding down Alt as you make a cut allows you to create a series of equally sized divisions. This is a useful technique for glitch and stutter editing.

Topping-and-Tailing

Another important editing technique with audio regions is the so-called topping-and-tailing. This ensures that there are accurate and clean edits at the front and end of the audio passages, thus reducing any unnecessary noise and other unwanted artifacts. It also allows you to set a precise length for a region, which is vital in situations where you want it to loop effectively.

Regions can be manipulated easily and trimmed to fit by simply calling up the Resize Tool, which appears when you move the pointer to the bottom left- or right-hand extremes of any region. The pointer

Figure 5.11 Moving to the edge of a region will transform the Pointer Tool into the Resize Tool, allowing you to adjust the respective size of the region.

changes to an icon similar to that of a squared bracket with arrows on either side. With this tool enabled, simply drag the region to the desired length. As previously mentioned, this is a useful way of restoring an audio region to its original form if you've been slightly overzealous with your editing!

Using the Marquee Tool

A useful alternative to the Scissors Tool is Logic's Marquee Tool. Rather than cutting a region into two, the Marquee Tool allows you to select audio events within one or more regions, and then either delete the offending notes, or indeed, create a new region based on the Marquee's boundaries. Overall, the

Figure 5.12 Use the Marquee Tool to enclose a selection across one or more regions – this can then be separated into a new region, deleted, or muted.

Marquee Tool works best as a means of time-slipping notes, especially on a multitracked recording of drums – first, by selecting the offending hit in question using the Marquee Tool, separating the selection into a new region, and then moving the region/s accordingly.

Once you've selected the Marquee Tool, simply drag-enclose the area you want to select – whether it's a few notes within a region, for example, or a portion of time across multiple regions. You can add to or remove from the selection at any point by Shift-clicking on your selection – adding in another track, for example, or increasing the size of the selection. Remember that any positional information will be set using the current Snap setting, so if you need your edits to be particularly precise, you'll either need to zoom in on your edit or use the control modifier to temporarily disable any snapping.

Figure 5.13 The Marquee Tool is a useful way of muting audio events within a region – simply enclose the area and press Ctrl + M.

Once selected, you can simply hit the Delete key to remove the offending selection – leaving two regions (and an empty space) on either side of an edit. Alternatively, press Mute (Ctrl + M) to silence the selection but leave the region in place. To divide the selection into a separate region, simply Ctrl-click on the Marquee selection, and then Logic will divide the original region into three new regions.

Snap to Transients

A useful extension of the Marquee functionality is the ability to extend the selection to transient hits within the region. This is a particularly good way of editing drums, allowing you to accurately define a Marquee selection to hits within the region – this would allow you to neatly separate out a two-bar drum loop, for example, precisely slicing out an individual drum hit. Start by making the Marquee selection over the broad area you want to work with. Next, move the start or end points of the selection using either the left and right cursor keys to move the out-point of the enclosure or the left and right cursor keys with the Shift key held down to reposition the in-point of the enclosure.

5.8 Fades and Crossfades

Fades and crossfades are an important part of audio editing on many different levels – whether it's just top-and-tailing an audio file, for example, or splicing together two adjacent audio files without creating

Figure 5.14
Use the Crossfade Tool as a simple, graphical way of fading a region in or out.

any unnecessary clicks, or transitional discontinuities, between the two regions. In Logic, there're three different ways to create fades and crossfades, each of which has its own strengths and weaknesses.

Using the Crossfade Tool to Create Fades

Probably the most immediate way of applying a fade is to use the dedicated Crossfade Tool, which can be used to create both fades and crossfades. The advantage here is that the tool works in a graphical way, with the length of the fade established by the proportion of the region you enclose using the Crossfade Tool. This makes it easy to establish the fade in relation to the waveform – maybe fading a region out, for example, as the sound decays away.

To create a fade at the start or end of an audio region, simply drag the Crossfade Tool over the appropriate segment you want to fade. Logic will then update the region, displaying a shaded area to indicate the length of the fade. To change the fade at this point, simply drag the small vertical white line accordingly, moving it backward or forward to lengthen or shorten the fade. Clicking on the fade itself allows you to adjust the relative curve of the fade – moving between a linear fade (the default) and the exponential and logarithmic curves.

Where two regions are "butt edited" up against one another, you might want to consider placing a crossfade. A crossfade is designed to smooth out potential discrepancies between the two regions, so that you're left with a smooth transition between the two components. For example, even if the two regions are related in some way, the process of butting one up against the other will result in a small click as the waveform jumps from one position to another. In worst-case scenarios, it might be that the two takes are radically different from one another, possibly necessitating a longer crossfade so that the edit isn't noticeable.

Figure 5.15
With the Crossfade Tool active, click on the white fade line to change its curvature or length accordingly.

You can create a crossfade by dragging the Crossfade Tool across two adjoining regions, with the respective length of the crossfade dictated by the size of the enclosure you make. Of course, in the case of a discrete butt edit, it might be that only a small crossfade is required, although this is usually best placed just ahead of a beat so as not to damage any transient information. Where the discontinuities are greater, though (for example, maybe the adjoining notes have slightly different tuning), you'll need to consider using a longer crossfade.

Changing the Drag Mode to Create Fades

Along with the various Snap modes, Logic also includes several different Drag options, which can be found in the top right-hand corner of the Tracks area. The Drag mode defines how two regions interact when they are dragged over one another.

By default, regions are set to the Overlap drag mode, meaning that the regions can sit on top of one another, but the latter part will always take priority and mute the region that lies underneath it. This is a logical starting place for editing, especially when you might want to slip back and forth the crossover points between the two regions so that they fit appropriately. No Overlap, however, simply means that the first region will be edited in size to fit.

Figure 5.16
The drag mode defines how Logic will behave when two regions are placed on top of one another.

As an alternative to overlapping regions, Logic can also employ an automatic Crossfade Tool within the Drag menu, known as X-Fade. In X-Fade mode, the process of moving or resizing a part over another audio region will create a correspondingly sized crossfade. Therefore, by careful use of the region resize tool, you can quickly produce perfect crossfade edits between multiple regions – initially "butting" the edit to a bar, and then, resizing the inserted region/s back to create the crossfade.

Figure 5.17 With the X-fade mode engaged, regions that overlap (either by resizing or being moved) will be crossfaded accordingly.

Using the Fade Parameter to Create Fades

The final method for applying fades is to use the Fade feature as part of the Inspector's Region Parameters. Given the intuitive methods of applying fades that we've already seen, you'd be forgiven for wondering why you'd bother using the Inspector method. The first significant advantage with this approach, though, is the ability to create incredibly short fades without zooming in to the audio region/s in question. Second, it also provides a convenient solution where you need to apply fades to multiple regions at once – something that is impossible to achieve using the Crossfade Tool.

Figure 5.18 The Inspector offers a precise way of entering in fade information for a given region.

5.9 Speed Fades

As an extension to the fade parameter found in the Inspector, it's also possible to create a Speed Fade. Technically, a Speed Fade isn't something you'd want to apply in conventional audio-editing activities, although as a creative effect, it's a joy to behold! In theory, a Speed Fade approximates how a DJ might "spin in" a record, as well as slowing it down at the end to create a dramatic tail-off. Therefore, rather than modifying the amplitude of a region, a Speed Fade modifies its speed – pitching the region up across a "fade in," while slowing it down over a "fade out."

You can create a Speed Fade in much the same way as you create an existing fade. It probably makes most sense, therefore, to start by creating a traditional fade out – using the Crossfade Tool – over the area you want to Speed Fade. Once this has been established, turn to the Inspector and change the fade mode from Fade In or Fade Out to Speed Up or Speed Down, respectively. Logic should now apply the fade, but in the form of a Speed Fade. Any subsequent changes to the length of the fade, therefore, will now change the duration of the speed-changing effect.

Figure 5.19 Use the Inspector to change the fade type – this allows you to create a distinctive Speed Up or Speed Down fade.

5.10 Introducing the Audio Track Editor

In addition to the editing techniques we've described within the Tracks area, it's also worth noting the presence of the Audio Track Editor. The Audio Track Editor is somewhat akin to the Piano Roll Editor's role with MIDI, in that a double-clicked region will open the editor window at the bottom screen. Functionally speaking, much of what can be achieved on the Tracks area level can be duplicated here, although as we'll see, there're two further levels of functionality that expand on this basic premise.

Figure 5.20 The Audio Track Editor replicates many of the features of audio editing in the Tracks area, but will become more important as we move into areas like Flex Time.

The first point to note is Flex Time and Flex Pitch, both of which are discussed in 5.15 and 5.20, respectively. Flex Time and Flex Pitch offer a degree of pitch and time-based elasticity with your audio, allowing you to change the timing of a bass guitar track, for example, or fix intonation problems on a vocal. With both Flex Time and Flex Pitch it's possible to perform minor corrections from the Tracks area but if you want more detailed control (especially in relation to Flex Pitch) you'll need to use the Audio Track Editor.

The other important point to note is that the Audio Track Editor can also take us down a level to the Audio File Editor. On the surface of it, the Audio File Editor looks very similar to the Audio Track Editor, although its role and application is different. In essence, the difference between the Track and File editor is the concept of a "destructive" edit. With the Audio Track Editor all edits are nondestructive, meaning that the original recording data means unmodified. So, even if you change the pitch or the timing of the audio file it remains as it was originally recorded. The Audio File Editor, however, modifies the original data, thereby making it a "destructive" audio editor.

5.11   Quick Swipe Comping

In Chapter 4, we introduced the notion of cycle recording and Logic's ability to keep these multiple performances – or takes – as part of a Take Folder. This is certainly an effective way of managing multiple recordings of the same performance, reducing potential clutter on screen, and allowing you to group related recordings in a logical way. However, once you've recorded all these different takes, you then need to decide which takes you want to use, potentially switching between several different takes as part of the same finished "performance."

Technically speaking, the process of editing between different takes is known as comping, and Logic includes a dedicated feature specifically for this task – Quick Swipe Comping. Using this feature, you can quickly audition different takes, as well as dynamically switching between them without having to use complicated editing tools and crossfades. You also have the ability to create several different comps that allow you to experiment with different treatments of the same musical part without damaging any existing comp decision you might have already made.

5.12   Expanding the Take Folder and Creating Basic Comps

You can expand a Take Folder to see its contents by clicking on the small arrow in the top left-hand corner. You should now be able to see the various takes, alongside shading to illustrate which take is currently being used. At this stage, you could audition each take – either by using the small drop-down menu in the top right-hand corner of the Take Folder or by simply clicking on each respective take.

The idea of Quick Swipe Comping is that you can quickly move between the different takes – "swiping" the word, phase, or series of notes that you want to include. Adapting the quick swipe edit points between the different takes is as simple as dragging over the respective take to change the amount of highlighted audio. Note how changes on one take are reflected in another (only one take can play at any point in time), and how the top line displays the current edit in its complete state.

Figure 5.21 Once you've opened your Take Folder, you can quickly start to piece together a preferred Comp between the various takes.

Comps that you have made can be named using the drop-down Quick Swipe Comping menu, which is accessible to the right of the small arrow used to expand the Take Folder. Note that it's also possible to Duplicate and Delete multiple different Comps, allowing you to try out several different editing strategies, and then audition the relative merits of each, by switching between the different completed Comps (which, again, are selectable from the Quick Swipe Comping drop-down menu).

Figure 5.22 Comps can be named, duplicated, and deleted, allowing you to explore multiple different versions of the same line.

5.13   Time Slipping a Quick Swipe Comp

So far, we've seen how we can use the Quick Swipe Comping system as a means of switching between different takes of the same musical part. However, what we haven't yet addressed is the need to slip potential edit decisions that you've made in time – of course, you can always move the whole Take Folder, but what if your concerns are with only one or two notes within one of the takes? If you attempt to move a region at this stage, you won't get very far – either because of the inability to use the Scissors Tool or that any attempt to move the region results in the Comp itself being changed.

The solution to this problem is the ability to switch the Take Folder's Quick Swipe Comping functionality on or off accordingly. You can see whether Quick Swipe Comping is enabled by looking at the top left-hand corner of the Take Folder, just to the right of the Take Folder's drop-down menu. By disabling the

Quick Swipe Comping, you'll find that Logic's original toolset – namely, the Scissors Tool and Marquee Tool – now work as expected. Given that the broad comping is finalized (this time slipping is much easier to achieve after you've swiped the takes!), you can go about slicing out the offending notes and then repositioning them.

As we'll see later on, though, this isn't the only solution for time-slipping material within a comp.

Figure 5.23 By disabling the Take Folder's Quick Swipe functionality (by clicking on the small icon on the top left-hand corner of the Take Folder), you can use the standard editing tools, particularly with respect to time-slipping elements of the performance.

5.14   Flattening or Merging a Comp

Finished, or even partly finished, comps can be handled in a number of different ways. First of these is the option to flatten the edits you've made – "unpacking" the Take Folder, and creating a single track with all the edits in place. Flattening has its advantages and disadvantages – on the negative side, you can't return to the edits in such an intuitive way as the Quick Swipe Comping feature, but on the positive side, you'll gain a "clean" edit of your chosen comp, complete with editable crossfades within each edit point.

You can choose to flatten your edits through the Take Folder menu, with the resultant output displayed as a series of regions with crossfades between them. Flattened in this way, you can easily carry out any timing modifications (using an alternative approach to what we've seen previously), as well as adjusting the default crossfade times using the same processes that we've explored earlier on in the chapter.

Figure 5.24 Flattening your comp will create a single track with all the various takes edited accordingly.

A more destructive version of the Flatten function is the Flatten and Merge option. Using Flatten and Merge will create a new file, complete with the edit points and crossfades. If you're completely happy with your edits, Flatten and Merge could be an appropriate choice, especially if you want a more transportable version of the track to import into another DAW. Unless you choose to specifically delete the original takes, you'll always have the option of reimporting the files from the Project Audio Browser into the Tracks area should you decide to revisit what you've done at a later point.

Figure 5.25 Flatten and Merge is a more destructive way of finalizing your comp – with a completely new audio file created from your editing decisions.

Whether you're carrying out a Quick Swipe Comp or Flattening a Comp Logic will work with the current default Crossfade Time and Curve settings established under the General tab of Logic Pro X > Preferences > Audio. In most cases, the default setting of 20 ms is sufficient, although for particularly taxing comps, it might be the case that this has to be raised or lowered accordingly. Arguably, though, such fade adjustments are best applied to a flattened comp and this preference left as is.

Figure 5.26 Use the General tab, found under Logic Pro X > Preferences > Audio, to change the default crossfade time for Quick Swipe Comping.

Unpacking a Comp

As an alternative to flattening your edits, it's also possible to unpack the Take Folder, placing all the individual takes, as well as the various comps (complete with crossfades), onto separate tracks. Unpacking a Take Folder is a messy option in terms of the Tracks area and the amount of tracks it can potentially use up, but it does offer the best solution in terms of complete audio flexibility. For example, an unpacked take can be easily moved to new tracks – maybe using some outtakes from a main vocal as backing vocals. Alternatively, it might also be that the remit of editing goes above and beyond what can be achieved in Quick Swipe Comping alone – like a vocal requiring detailed pitch correction in Melodyne, for example, or elements of time compression and expansion.

Figure 5.27 Unpacking a Take Folder will explode all the original takes onto separate tracks, as well as flattening any comps you've made along the way.

As the reverse of unpacking Take Folders, you can also use the contextual menu function Folder > Pack Take Folders (available by ctrl-clicking on a selection of regions) to combine several existing audio files into the same folder. The option to pack Take Folders potentially solves the previously suggested problem of adjusting timing issues – simply unpack the folder, move the required take, and then repack accordingly. The feature can also be an interesting way of creatively switching between a number of concurrent drum loops – packing them into a shared Take Folder, and then using the Quick Swipe Comping to create complicated jumps between the loops.

5.15   Flex Time Editing

So far in our exploration of production process in Logic, we've looked at the "conventional" tools and techniques of audio editing – namely, the ability to slice, resize, and reposition audio regions. However, in the past 5 years or so, musicians and producers have grown increasingly used to having a greater degree of "elasticity" with their audio – either using specific third-party applications like Melodyne, for example, or an increasing amount of tempo and time-adjustable functionality directly embedded into a DAW of choice. Therefore, having already seen some of the tempo-elastic possibilities of Apple Loops, Flex Time editing transforms the potential flexibility of audio regions within your project.

Why Use Flex Time?

As the name suggests, the principle use of Flex Time is to modify the timing properties of an audio region and the "events" contained within it. Of course, using conventional editing techniques, we can always use region slicing to create such timing moves, but this can often be a laborious task necessitating multiple slices and complicated crossfading to make an edit sound transparent enough to the listener's ear. What's next to impossible to achieve using standard editing techniques, though, is the ability to change a note's duration, whether it's shortening or lengthening a note to optimize its phrasing.

The first application of Flex Time, therefore, is the ability to move notes backward and forward within a given region – achieving everything from the ability to "slip" a note so that it plays more in time with events in other audio regions, through to the ability to completely rephrase a melody line (including, as we've seen, the duration of the notes) so that each note falls in a different position in the bar. As the audio is "intelligently" time stretched (using the currently selected Flex Time mode) no additional editing is required, and the transition between the edited and unedited notes remains smooth and glitch-free.

Another important application of Flex Time editing is the ability to quantize audio events within a region – making a bass play exactly in time with a drum loop, for example, or changing its groove from a straight 16th feel to a bassline with a subtle "swing" feel. Again, this is something that's possible to achieve with conventional editing but can be achieved in Flex Time in a matter of seconds.

5.16   Flex View and Flex Time Modes

To start working with Flex Time in Logic, you'll need to engage the Flex View mode using the menu option Edit > Show Flex Pitch/Time, or by clicking on the Flex View icon in the left-hand corner of the Tracks area. Once the Flex View mode has been engaged, you see some extra information in each of the track headers, allowing you to select a unique Flex mode for each track in your project. Ideally, you should try to match the Flex Time mode to the type of track being handled. The modes are largely self-explanatory, but let's take a look at their principle application in music production.

Figure 5.28
To begin working with Flex Time, you'll need to engage the Flex View mode (Edit > Show Flex Pitch/Time).

Slicing

The Slicing mode is the obvious choice to use with drums and percussion sounds and makes a point of preserving transient details in the original recordings. Behind the scenes, the audio region is "flexed" using slices, using much the same process as we've seen with manual editing.

Rhythmic

The Rhythmic mode is a good "catch all" solution for a range of audio files that contain both pitched and percussion elements – this might include a drum loop with some in-built bass parts, for example, or a rhythm guitar. Some "stretching" artifacts might be evident under closer scrutiny (or extreme settings), either as part of the body of the sound or the transient details.

Monophonic

If you are working with a "melodic" instrument that doesn't play chords – like bass, vocals, or lead guitar – then opt for the Monophonic Flex mode. This provides a good balance between transparency and CPU usage.

Polyphonic

The Polyphonic mode is arguably the most CPU-intensive Flex mode and should be reserved for polyphonic instruments, or indeed, complex submixes, that require high-quality time stretching with any unnecessary artifacts kept to a minimum. Remember that you can always bounce the finished Flex Time edit (for more information, see Section 5.19).

Tempophone

Definitely an inappropriate choice for smooth "flexing," Tempophone is a Flex mode deliberately designed to sound obtrusive and noticeably modified. Use Tempophone when you want to use Flex Time almost like an effect, with the sound appearing increasingly grainy and stretched as you move increasingly further away from its original duration.

Speed

As the name suggests, the Speed Flex mode modifies the speed of an audio event to account for changes in time. Compacting the placement of notes, for example, will speed up the playback, whereas stretching out notes will slow it down. This is much the same as modifying the pitch and tempo of track using a turntable's varispeed control, or indeed, modifying the playback speed of tape. The Speed Flex mode, therefore, is best reserved as a creative effect, where you want audio to sound deliberately repitched accordingly.

Besides engaging the Flex mode for the track, it's also worth to check that the Flex functionality has been engaged as part of Inspector's Region Parameters box. In most cases, this will already be activated, but occasionally a region can have its Flex functionality removed.

Figure 5.29
A variety of different Flex modes allow you to optimize the time-correction algorithms for the type of audio material you're working with.

5.17   Transient Markers and Flex Markers

Flex Time makes use of two types of markers – Transient Markers and Flex Markers – to carry out the "flexing" process. Transient Markers are first used to divide up the audio events within a region – creating markers for each word on a vocal take, or the individual hits within a drum performance. Flex Markers, however, identify the notes you wish to modify and how the notes are handled on either side of the Flex edit. Although this sounds complicated at first, their use and application soon becomes apparent once you start using Flex Time.

Creating Transient Markers

Whenever you engage a Flex Time mode, Logic will go through the regions on the selected track and assign a series of Transient Markers. This is a relatively quick process, and once finished, you'll see a series of small gray hairlines to indicate their placement. Note that with Take Folders, though, the markers aren't evident on the "top" level but can be viewed and modified once the Take Folder has been expanded.

Flexing Multiple Audio Events

With a series of Transient Markers placed, you can now start to explore various applications of the Flex Time technology. Probably the first and most immediate application is to resize a region and its contents accordingly. Previously, of course, the act of resizing a region would result in the "edit window" being resized, rather than modifying any of its contents. In Flex Time mode, though, the events will be expanded or contracted accordingly. This makes it easy, for example, to conform a drum loop imported in at one tempo to your current project's tempo, simply be resizing to the corresponding region end point.

Note that the expansion or contraction of any audio information within the region is always color-coded accordingly. Audio events that have been contracted will take on a darker hue, whereas those that have been expanded will take on a lighter shade. Where the stretch is too extreme, the shading changes to a red color, suggesting you try an alternative solution to the problem at hand!

Figure 5.30 Once a region has been placed in Flex mode, any attempt to resize it will result in the audio being expanded or contracted accordingly.

Besides changing the region as a whole, Flex can also modify notes within it using a series of one or more Flex Markers. The art in understanding and using the Flex Marker system lies in the two contrasting ways Flex Markers can be placed, as indicated by the changing appearance of the Flex Marker Tool that appears as you float over the region in question.

By clicking in the upper half of our waveform display, we can create a single Flex Marker, which is set to the nearest Transient Marker and effectively acts as an "anchor" to the flexing process. By moving the Flex Marker in either direction, audio will be expanded or contracted on either side of the marker, again using green and orange color coding to indicate the qualities of expansion and contraction taking place.

Figure 5.31 A single Flex Marker will expand or contract the audio events on either side of it, as indicated by the lighter and darker shading.

Flexing Single Notes

A far more musical way of using Flex Time is to use three adjacent Flex Markers to move a single note backward or forward accordingly. Although the process of creating multiple Flex Markers might sound complicated, it's easy enough to execute. With the Flex Time view active, move the pointer to the lower half of the region and note how the icon changes from one marker to three. Now, click on the note you want to move and note how three markers are created – one preceding your editing point, one on your edit point, and one succeeding your edit point. Moving the central Flex Marker moves the desired note position but preserves the timing on either side of the edit.

Figure 5.32
Placing three Flex Markers allows you to move a note's position without affecting the timing of notes on either side of it.

Another interesting way you can apply Flex Time is to tighten notes between adjacent regions – maybe aligning a note on a bass guitar, for example, to precisely play in time with a hit on the kick drum. First, you'll need to ensure that both tracks have been assigned a Flex mode, and the track that you want to lock the note (or notes) to has been shuffled up or down the tracklist so that the two sit adjacent to one another. Again, we start to move our selected note using the "three marker" icon found in the lower half of the waveform display. This time, however, drag the mouse up or down the adjacent track and note how the Flex Time marker snaps to transients on this track (the positioning line should also change to yellow to indicate this is happening).

Figure 5.33
By moving to an adjacent track lane, you can snap a note so that it aligns with another part.

Figure 5.34 Click on the small X to delete a Flex Marker.

Deleting Flex Markers

In keeping with the nondestructive ethos of editing audio in Logic, it's also worth noting that Flex Time edits can be undone at any point in the production process, returning the audio region back to its original state. One gradual way of achieving this is to delete individual Flex Markers, thereby removing any unwanted Flex Time edits, but retaining elements that you are happy with. To delete an unwanted Flex Marker, therefore, click on the small X icon at the top of the region.

When multiple Flex Markers have been assigned and you want them all to be removed, you can always Ctrl-click anywhere else on the region and select Reset All Flex Edits. This will restore the region to its original state, although subsequent regions on the track will retain any Flex Time edits that you have made. Therefore, to disable any Flex Time editing across the whole track, you can always switch the Flex mode back to its "off" setting to completely disable Flex Time editing.

Figure 5.35 As well as deleting individual Flex Markers, you can also delete a series of Flex Markers assigned to a region.

5.18   Quantizing Audio with Flex Time

One of the standout features of Flex Time has to be the ability to quantize the events within an audio region. Quantizing is the process of "tightening" a performance so that notes within fall more closely to the divisions of a bar – something that up until the introduction of Flex Time had largely been the preserve of MIDI editing (discussed in detail in Section 6.8).

To quantize the events in an audio region, you first need to select the region/s you want to quantize, and then pick an accompanying quantize setting from the Inspector's Region Parameters box. There're various settings to choose from, although in most situations you'll want to select the 1/16 setting, which places each note to the closest 16th (or semiquaver) division of the bar. Once selected, you'll notice that each note within the region has been moved slightly (unless the part was played exactly to the beat), as indicated by the orange and green shading.

From this basic quantize setting, we can then adjust various settings to change the feel and strength of the quantizing applied. For the sake of brevity, though, turn to Section 6.8 to understand fully what can be achieved here.

Figure 5.36 Flex Time allows you to quantize the performance of an audio region – shifting each respective note so that it precisely aligns to a rigid tempo grid.

Groove Templates between Regions

A handy extension of the basic audio-quantizing facilities using Flex Time is the ability to copy the groove of one audio region to another. In many ways, this can be a more suitable way of improving the timing between different audio regions – allowing two different parts to play tightly against one another, but not losing the feel of the song to a rigid 1/16th grid.

The first part of the process is to extract a groove template from a suitable source region – in other words, the part you consider to be "in time." To extract the groove, the track will need to be placed into an appropriate Flex mode (slicing is probably best) so as to create a series of Transient Markers. Once this has been done, highlight the region, and then select Make Groove Template from the region Inspector's quantize menu. Once the groove has been created, you can choose to take the track out of its current Flex Time, assuming you have no need for it to be "flexed" in any way.

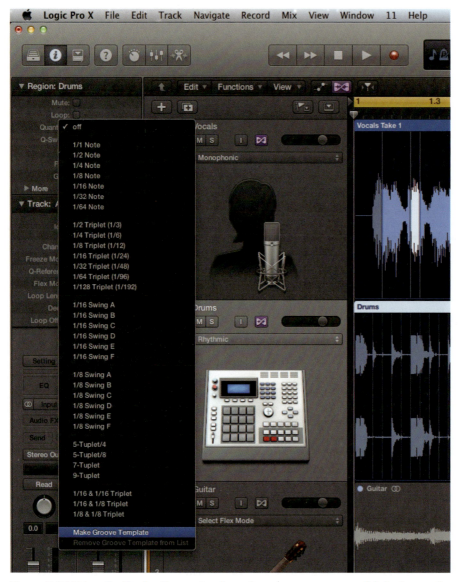

Figure 5.37 Using the Region Parameter drop-down menu, you can add the "groove" of an existing region to your palette of quantize options.

Applying the newly created Groove Template is much the same as the example of audio quantizing we've explored earlier on in the Flex Time editing section. In this case, however, you'll need to pick the particular Groove setting (rather than 1/16th, for example), which should appear at the bottom of the Quantize drop-down menu.

Figure 5.38 Having created our new groove, we can then apply it to another region in our song. The two parts should now play together perfectly.

5.19 Separating Regions Based on Transient Markers

One final feature made possible by Flex Time is the ability to divide a single region into multiple regions based on Transient Markers within it. In the case of a recording of a drum kit, for example, this could allow you to create individual regions for each drum hit. Given the flexibility of Flex Time's slicing mode (which achieves similar functionality in a more transparent way), the wisdom of applying this is somewhat dubious, although you might find some benefits in having regions separated in this way. One useful application, though, would be to divide a recording up into a series of "one shot" audio files ready to be imported into a software sampler like the EXS24.

You can slice a region that has been placed into a Flex Time mode by Ctrl-clicking on the region and selecting Slice at Transient Markers from the drop-down menu. Of course, once sliced, you can always take the track out of Flex Time mode to avoid any unwanted CPU activity. Should you want the regions to become individual sound files in their own right, then highlight them, and select Convert > Convert Region to New Audio Files from the region's contextual menu.

Figure 5.39
Once Transient Markers have been placed, you can separate a single region into a series of regions using the Slice at Transient Markers feature.

5.20   Flex Pitch

In addition to working with the timing properties of an audio file, you can also adjust properties in relation to Pitch. The important point to note is that Flex Pitch's assumed application is with monophonic tracks – in other words, musical parts that use a single melodic line, rather than a collection of notes to form a chord. As such, Flex Pitch is a great tool to use with vocals, flutes, lead guitars, and so on, but slightly less than useful with string sections, acoustic pianos, strummed rhythm guitars, or any other "polyphonic" audio.

Enabling Flex Pitch

Enabling Flex Pitch is much the same process as Flex Time – simply enable the Flex View (Edit > Show Flex Pitch/Time or press the Hide/Show Flex icon on the top of the Tracks area) and place the desired track into Flex Pitch mode. Unlike Flex Time, though, you'll want to make sure that Audio Track Editor is open to appreciate the fullness of what you can achieve. Even so, the Flex View in the Tracks area provides some indication of the part's intonation, as well as some basic editing options via the contextual menu, which we'll discuss in Section 5.21.

Figure 5.40
With Flex Pitch enable in the Tracks area, you have some indication as to the relative intonation of the part.

How to "Read" the Flex Pitch View

The Flex Pitch view is quite unique in that it provides a combination of a waveform overview, much like a traditional audio editor, as well as a piano-roll-like editor that provides the means of viewing and editing the region's pitch information. The pitch of the notes is displayed much like MIDI notes in the Piano Roll editor, illustrated as rectangular objects along the track lane. Of course, you can assess the relative intonation of each word using the position relative to the piano along the left-hand side of the Audio Track Editor. Where a note is slightly deviant in respect to its pitch, you'll see that the blob is not wholly shaded, suggesting that the pitch is either slightly flat or sharp.

Within each note, you'll also see a thin line that is used to indicate both the pitch drift and vibrato in the note. Many vocalists, for example, will slide up to a note or fall off the end of a note to create some level of emotion and expression to the part. Likewise, vibrato within the note will mean that the note fluctuates around a given pitch, rather than staying fixed at a precise frequency. Of course, there's balance to be struck between expression and a pitch-perfect performance, which is why it's important to see (and edit) these small movements and variations in pitch.

Figure 5.41
The Audio Track Editor provides a detailed overview of the Flex Pitch editing feature.

5.21   Typical Flex Pitch Edits

Clicking on a note within the Audio Track Editor reveals the different ways a note can be edited – Pitch Drift, Fine Pitch, Gain, Vibrato, and Formants – as indicated by the six nodes positioned around the note. Clicking on any of these nodes, or the note itself, allows you to edit the performance accordingly.

Basic Vocal Intonation

Fixing the intonation of a vocal performance, or any other monophonic pitched instrument for that matter, is a relatively straightforward exercise, thanks to the contextual menu. Ctrl-click in the Audio Track Area away from one of the note "blobs" and select "Set all to Perfect Pitch" from the menu that appears. This will now automatically correct the entire vocal, so that the intonation is perfect. If you don't like the results, simply pick "Set all to Original Pitch" and the performance will be restored to its original form.

In the Tracks area, note how the region displays a simplified version of Flex Pitch overview, although the display is orientated toward showing intonation drift rather than relative pitch. The important point to note is that the same contextual menu (with the "Set all to Perfect Pitch" option) is also accessed from the Tracks area level, making it quickly fixable from the arrangement.

As an alternative, you also have the option of manually adjusting each note one-by-one. This might be useful if the intonation is generally fine, but just mildy problematic on the occasional note. To adjust the fine pitch, therefore, click on the middle node on the top of the note and drag it up or down accordingly.

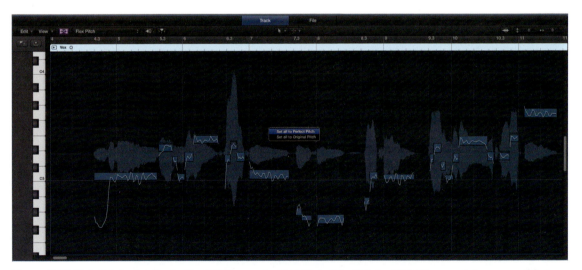

Figure 5.42 Ctrl-click in the Audio Track Editor and select "Set all to Original Pitch" as a quick means of fixing intonation.

Changing Note Pitch

You can change the coarse tuning of a note in semitones by clicking on the whole note rather than the editing nodes at the corners. Changing the coarse pitch is interesting as a means of repitching a melody for creative ends. Alternatively, duplicate the vocal, and you could also use the same technique as means of creating a vocal harmony track, with the vocal up a third (or more) from the main line.

Figure 5.43 Click and drag on the body of the note to change its coarse pitch.

Controlling Vibrato and Pitch Drift

Within the note, of course, we have both vibrato and pitch drift at either end of the note. Use the middle node at the bottom of the note to change the amount of vibrato – either increasing it; to add more

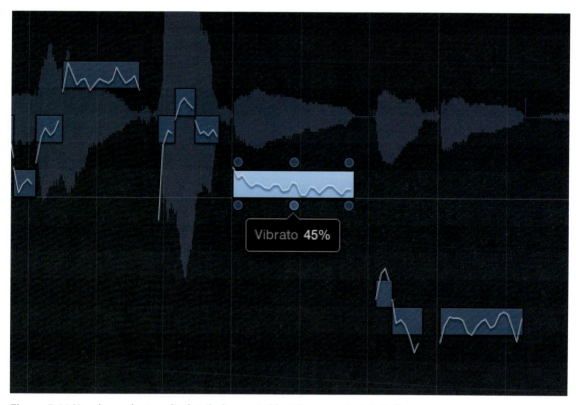

Figure 5.44 Use the nodes to edit detailed aspects like Vibrato and Pitch Drift.

dramatic intensity, or decreasing it; to make the note less dramatic. The two nodes at either end of the top of the note can be used to address slightly over ambitious pitch drift.

Editing Gain and Formants

Two of the more intriguing nodes are the Gain Control, found in the bottom left-hand corner, and Formant Shift found in the bottom right-hand corner. The Gain change is an interesting alternative to compression, especially on a vocal, where you might want to raise the level on certain words or notes that are otherwise getting lost in the mix. Changes in Gain are automatically reflected in the waveform view so you can immediately see the effect you're having.

Formants are frequency peaks in the sound spectrum that form a large part of the identity of the voice and stay at the same fixed frequency irrespective of the pitch of the vocal. Shifting both the pitch and formants of the voice, therefore, can appear to make the voice take on a type of "'Mickey Mouse" tone. Changing the formant, therefore, can be an interesting effect, especially if you want to subtly shift the gender of the vocals.

Figure 5.45 Changing the Gain and Formants offers precise control over instruments and vocals in Flex Pitch mode.

5.22 Freezing a Flex Time Track to Save CPU

Make no mistake about it – an extensive use of Flex Time editing can have a big hit on your CPU resources, especially if you use some of the more CPU-intensive Flex modes like the Polyphonic option. If you feel happy about the Flex Time edits you've made, therefore, you might want to consider "freezing" the track. Freezing is a common process in Logic (also discussed in Chapter 8, KNOWLEDGEBASE 3, Saving CPU: Freezing and More . . .) whereby an "off-line" rendered version of the track is created (complete with any plug-in settings or Flex Time adjustments), which is then streamed from disk in much the same way as an audio file. Although frozen, the track looks exactly the same – even retaining its Flex Markers – although any subsequent changes will result in the track needing to be re-rendered, which can take a few seconds to complete.

To freeze a track, you'll need to click on the Freeze icon as part of the track header. If you can't see this, select Track > Configure Track Header, and make sure the Freeze icon is checked accordingly. Before you initiate the freeze, another important point to check is the current Freeze mode, which can be established as part of the Track Parameters box, just below the Region Parameters box in the Inspector. Ideally, it should be set to Source Only, so that only the Flex Time edits are rendered in the form of an offline file. By comparison, the Pre Fader option renders both the source information and plug-ins – which isn't so appropriate as we haven't got around to mixing our track yet!

Figure 5.46
Using the Source
Only and Freeze
modes allows you
to render your
Flex Time edits
as an offline file,
making a significant
saving on your CPU
resources!

5.23 Editing Multiple Tracks Using Edit Groups

Most of our edits so far have involved single-tracked instrumentation – that is to say, an instrument that's recorded to a single track in either mono or stereo. However, there are plenty of examples where you might want to record more than one track on a single "pass": for example, recording a drum kit; an orchestra using a combination of close microphones, room microphones, and a stereo pair; or a guitar amp using an array of different microphones placed around the cabinet. In all of these examples, your edits will want to be achieved en masse, rather than having to individually slice each accompanying region every time you make an edit.

Creating Edit Groups

Logic's Grouping function allows you to link together tracks in respect to both editing and mixing. In Chapter 8, we're going to be looking more closely at the applications of groups in mixing, but for now, we'll concentrate solely on their use in editing.

You can assign one of more tracks into an edit group through the Mixer area, which also provides a current overview of the grouping status. Each channel strip on the Mixer page has its own Group Slot, which you should be able to see just below the output assignment. Therefore, to assign a channel strip to group, simply click on this box and select a corresponding group from the drop-down menu. Note that you can also mass-enable several channels strips by Shift-clicking to augment your existing selection, and then selecting an appropriate group using one of the channels strip's group slots.

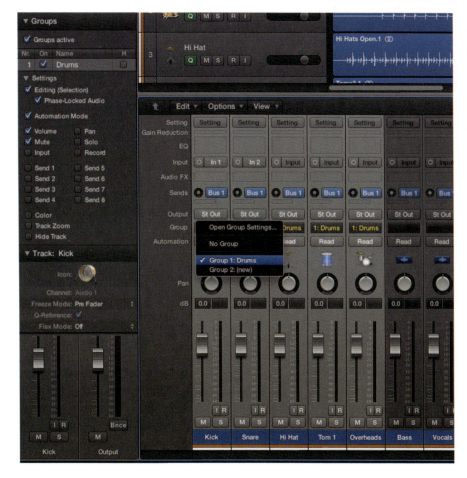

Figure 5.47
Create an edit group using the Mixer, ensuring you select the Editing (Selection) and Phase-Locked Audio check boxes.

Whenever you place a channel strip into a group, Logic will open the Group Settings window as part of the Inspector, which allows you to define how each group works. The principal option we're interested in at this point is the Editing (Selection) and Phase-Locked Audio check boxes, as this will facilitate our edits being simultaneously carried out across all the regions. If you can see these parameters, make sure you click on the arrow next to the Settings text to fully expand the group settings.

Once the group has been established, you can go about performing your edits using any of the techniques we've described – using basic editing features like the Scissors and Marquee Tools to perform rudimentary edits, or stepping up to features like Flex Time, where your edits need to be more detailed. One point to remember, though, is the choice of "master" region you use to perform your primary edits. In the case of a drum kit, for example, the Overheads track makes most sense as this gives an accurate overview of all the notes played, rather than the individual notes that relate to just one part of the kit.

Figure 5.48 With the edit group active, all edits will be carried out across the associated regions. This is a powerful feature when used in combination with Flex Time editing.

Toggling Groups

Of course, once you've enabled a group, you're not restricted to all subsequent edits being carried out across all the regions. By toggling the groups – using the so-called Group Clutch key, which is assigned to Shift + G – you can disable the group functionality across all the active groups. This then allows you to carry out edits on individual tracks as appropriate, returning the group functionality (using the Group Clutch key) once the edits have been made. You can see whether or not groups are active by viewing the small group slot on the Mixer page – when the legend is yellow, the group is active, and when it's gray, the groups are currently disabled.

5.24   Drum Replacement

In addition to the use of edit groups and Flex Time to tighten a multitracked drum recording, you might want to turn to the Drum Replacement feature to either completely replace an individual drum sound within the kit recording or augment the existing sounds with a sample-based layer. Drum replacement is a common technique used in professional circles (although few actually admit it!), but has previously required additional software – like Drumagog – to achieve in Logic.

Most drum replacement tends to focus on the kick and snare drum, as this contains the majority of the sound energy behind the kit, but equally because it's relatively easy to achieve. By comparison – the overheads would be next to impossible to replace, mainly because of the number of different sounds (kick, snare, hi-hat, toms, and so on) occurring at the same time. In short, therefore, stick to tracks that have been spot microphoned to a single part of the kit, rather than tracks with more ambient microphone placements.

To start the replacement process, select the track that you want to replace or double, and choose the menu option Track > Replace or Double Drum Track. A dialog box allows you to refine the drum-replacement process using a few easy-to-use parameters. First, of course, you need to set the instrument type from the drop-down menu, which includes options for Kick, Snare, Tom, and Other. As you'd expect, the threshold control establishes the amount of triggers generated – if there's lots of background noise, for example, you might need to raise the threshold higher, otherwise pull it down to the point where all the drum hits have a yellow "trigger line" attached to them. You should be able to see all the triggers point on the accompanying regions.

Figure 5.49 Once you've chosen a track to work with, select Track > Drum Replacement/Doubling to initiate the drum-replacement process.

The Drum Replacement feature works in conjunction with the EXS24 sampler, and as part of the replacement process, you should have noted that an instance of the EXS24 has been loaded, alongside MIDI trigger region, and a default sample setting. Therefore, once you've established the threshold, you might want to press the Prelisten button to hear the replacement in action (both the replacement part and the original audio track should play together solo'ed out from the mix), as well as stepping through the accompanying library settings to audition different EXS24 drum sounds. Once you're happy with the result, hit the OK button to take you out of the Drum Replacement/Doubling dialog box, leaving the MIDI region and instance of the EXS24 in place.

Once you've created the basic trigger part, you can then spend plenty of time refining the results as you see fit. Of course, at any point you can change the selected EXS24 sample, especially as you get closer to the final mix and have a better idea of how the part should sound. It might also be relevant to take a closer look at the MIDI data that has been created – possibly deleting unwanted notes, for example, or changing the velocities to achieve the level of dynamics you require (we'll discuss MIDI editing in more detail in Chapter 6). As the sample is devoid of any spill, there's also plenty of potential to add reverb, compression, and EQ, without the fear of what might happen to the rest of the kit.

Figure 5.50 Using the Drum Replacement settings you can fine-tune the replacement process, adjusting the threshold and the type of sound you want to replace.

5.25 Audio File Editor Principles

Given the exhaustive editing possibilities within the Tracks area and Audio Track Editor, you could be forgiven for wondering why it was necessary to have another "layer" of audio editing within Logic. The Audio File Editor, though, adds some important additional features to work with audio in Logic universe, some of which might be less useful to your day-to-day workflow, but still essential in getting the most from Logic.

Rendered "Off-Line" Editing

The first important point to note about the Audio File Editor is the fact that a large number of processes carried out within it are destructive. Rather than simply modifying playback information, therefore, many of the changes involve rewriting the data that are contained within an audio file – a gain change, for example, permanently modifies the amplitude of the data itself, rather than simply playing back the file slightly louder or quieter. Likewise, a time stretch will permanently render a longer or shorter version of the file accordingly. This technique of permanently changing an audio file is often referred to as an offline edit.

Of course, actions taken in the Audio File Editor have a degree of "undo" available to them, but it's important to remember that decisions made in the Audio File Editor might irrevocably damage the audio files used in your project, not to mention the potential havoc caused by projects that share audio files among them. However, there're many examples of situations where you want to create a permanent change to the files, especially in situations where the audio file becomes the final "product" – like a sample ready to be imported into the EXS24 sampler, for example, or the final mix of a song ready to be burned onto an audio CD.

Therefore, as a good safety system, it's worth making copies of any audio regions you wish to transform using the Audio File Editor. This can be done by Ctrl-clicking on the appropriate region and choosing Convert > Convert to New Audio Files. A quick check in the Project Audio Browser should indicate the creation of a new audio file, which you can modify, safe in the knowledge that the source files remain untouched. As an alternative, the Audio File Editor also includes its own backup facility, found under the local menu option Audio File > Create Backup.

Figure 5.51 As the Audio File Editor is a destructive process, you might want to consider creating a backup of your audio region before you perform any edits. One way of doing this is via the contextual menu option, Convert > Convert Regions to New Audio Files.

Accuracy

Although it is possible to edit to a high degree of accuracy in the Tracks area, the Audio File Editor remains one of the best places to edit audio information down to individual sample frames. Generally speaking, this happens very rarely. One possible example, though, is the ability to use the Audio File Editor's pencil tool to draw out small clicks that can appear in an audio file (especially when it has been poorly edited without any crossfades!). You can also use the Audio File Editor to precisely align region start points so that they start exactly at the start of a note.

Figure 5.52 The Audio File Editor allows you to view audio data at its finest resolution – even down to the ability to "draw out" sample clicks and other unwanted information using the pencil tool.

Additional Functionality

In addition to being able to perform offline editing, there are also some useful "functional" additions that are primarily addressed in the Audio File Editor – from the ability to edit Transient Markers, for example, through to creating and positioning the so-called Anchor points. Again, these might not be so important to your day-to-day workflow in Logic, but they add to the repertoire of techniques at your disposal!

5.26   Audio File Editor Applications

Although not exhaustive, we've compiled some of the best applications and uses of Audio File Editor. There's plenty more to be found in this intriguing section of Logic, so it's worth exploring the Audio File Editor further as you grow more confident with your abilities.

Reversing Audio

This is one of my favorite applications of the Audio File Editor, and it is one of the most rewarding ways of "twisting" a given sound – from reverse cymbals, through to backward guitar solos! Start by creating a duplicate audio file using the Convert > Convert to New Audio Files menu command (otherwise, all occurrences will be reversed!). Once you've opened up the Audio File Editor, highlight the area you want to reverse (usually the complete sound file) and then select Functions > Reverse.

To position a reversed region, take a look at the "Marking Anchor Points" feature outline below.

Figure 5.53 Use the Audio File Editor's Reverse function to reverse a given sound – although remember this is a destructive offline edit, so it's best to make a copy of the region first.

Accurate Region Resizing

Use the Audio File Editor to change the in and out point of your regions, potentially negating any unwanted sounds that might precede the first note, for example, or stray sounds that occur after the last note. The current size of the region is indicated by the two markers at the bottom of the Audio File Editor, with an "in" and "out" point accordingly. Changing this won't change any relative positional information, so even if you modify the start point (to remove some noise, for example), your region will still play perfectly in time.

Marking Anchor Points

By default, all regions are positioned using their start points. However, it is possible to create an alternative anchor point positioned within the audio region. A good example of this in action is on reversed files, where the anchor point is probably best marked where the sound reaches its peak, rather than at the start of the region. You can place an anchor point by sliding the anchor icon at the bottom of the Audio File Editor. Now, when you move the region in the Tracks area, you see the new anchor point being used to place the region, rather than the region's start.

Cleaning Up Unwanted Noise

Small unwanted noises in-between notes can be quickly removed using the Audio File Editor. Simply drag-enclose the offending portion of audio and select either Silence or Change Gain from the Functions menu. So that there are no abrupt changes in the signal level, it might be prudent to opt for the Change Gain setting and pick a setting around −10 dB or so. This will attenuate the noise to the point that it is less noticeable but will not completely remove it. Again, remember these edits are destructive and will permanently change the source audio data!

It is also worth noting the other "audio hygiene" options available under the Functions menu – from the ability to normalize an audio file (in others word, lifting its amplitude to hit 0 dB FS), through tools like the Invert function, which can be used to permanently change the phase of an audio file.

Figure 5.54 You can use the Audio File Editor to accurately resize a region down to sample-level accuracy.

Figure 5.55 Anchor points are used as the "location reference point" for placing regions in the Tracks area.

Figure 5.56 The Audio File Editor's Functions menu contains a number of tools to modify the source data – from the ability to change the gain, through to phase inversion.

Time and Pitch Machine

Although we've seen a large amount of "tempo elasticity" courtesy of Flex Time and Apple Loops, it's also worth noting some of the "low-level" ways you can manipulate the pitch and timing of audio files using the Audio File Editor. Of course, before the introduction of Apple Loops and Flex Time, the Audio File Editor's Time and Pitch Machine was the only way a Logic user could manipulate the speed and pitch of an audio file, although there're still a number of applications that make this a valuable tool to have at your disposal.

As you'd expect, the biggest advantage of the Time and Pitch Machine is its relative accuracy – a time stretch, for example, can be as precise as a few samples, or in the example of a pitch transposition, as fine as a few cents. It's also notable as one of the most flexible ways of adjusting both pitch and time, primarily because it can move between the so-called free transposition, where the pitch and time are freely and independently adjustable, and "Classic" mode, which links together pitch and tempo/time changes, working in much the same way as a tape machine's varispeed control.

The Time and Pitch Machine can work either with the region as a whole (selecting Edit > Select All from within the Sample Editor) or within a selected time period within the region. For example, this functionality allows you to individually repitch notes within a vocal line without having to separate each note of the performance into a series of individual regions.

Once you've selected the area to be processed and opened up the Time and Pitch Machine (Functions > Time and Pitch machine), you can then adjust the relevant parameters in the accompanying dialog box. First of all, you'll need to set the mode – using Free mode to modify the pitch without modifying time (or the other way around), for example, or Classic mode when you want the feature to work much like varispeed (this can sound great on drums, for example). With tempo, you can either specify the change in tempo, a percentage, or the length of the file (either in samples, SMPTE, or Bars), while the amount of pitch change is measured in cents. You might also want to optimize the Algorithm for the material you're working with, especially if the Universal setting isn't delivering effective results.

Figure 5.57 The Time and Pitch machine provides one of the most flexible and accurate ways of modifying the tempo, duration, speed, and pitch of an audio file.

5.27   Inserting and Deleting Sections of Your Song

One inevitable task of producing a song in Logic is the need to add or remove a segment of time at some point in the project's arrangement – whether you need to remove an unwanted chorus, for example, or add several bars after the third chorus to form the middle eight. Of course, you could perform these tasks using some of the tools and technique we've discussed, or instead, you could turn to Logic's specific "Cut/Insert Time" feature to perform this even more quickly and easily!

To take advantage of the Cut/Insert Time feature, you'll first need to draw the locators at the top of the Tracks area to indicate the part of the song you want to work with – either the area that needs time "inserted" into it, or the portion of the song you want to remove. Once this is done, use the menu option Edit > Cut/Insert Time, and choose an appropriate selection (Cut, Insert, or Repeat).

Figure 5.58 Use the Cut/Insert Time feature to extend or shorten your arrangement accordingly.

5.28   Working with Tempo and Loops

Throughout this chapter, we've seen a number of principal ways in which we can deal with the tempo-based material in Logic, especially in relation to the application of Flex Time. However, there are a variety of problems that arise when dealing with tempo-based material, all of which necessitate a unique solution to the problem in hand – whether it's the ability to change your project's tempo based on the tempo of an imported loop, or being able to Beat Map your timeline to a performance that wasn't played to click.

Follow Tempo

Some, but not all, audio files will have tempo information embedded within them – this includes Apple Loops files (as we've already seen), but also material you record into Logic, and file formats like ReCycle's REX. In these situations, it's usually as simple as clicking on the Follow Tempo as part of the Region Parameters box so as to have the loop conform to your current project tempo.

Figure 5.59 If your audio file has tempo information embedded within it, you can use the Inspector's Follow Tempo option to have the region automatically conformed to the project's tempo.

Time Stretching a Region to Fit

If your audio file/loop doesn't include tempo information, there're a number of different ways to make it manually conform to your project tempo. First, of course, you can use the Flex Time functionality – engaging a Flex mode and then resizing the region so that it fits precisely to a number of bars. Another option is to use the Time Stretch option as part of the Edit menu. Once you've selected the region in question, pick either Edit > Time Stretch > Time Stretch Region Length to Nearest Bar or Edit > Time Stretch > Time Stretch Region Length to Locators, as appropriate. The option of stretching to Locators is useful if the amount of correction falls outside the "nearest bar." In this case, use the Locators to set the size of the eventual time stretched region.

Working Out the Tempo of a Region/Loop

Another option to manually conform the loop is to use Logic's Detect Tempo feature, which will then allow you to use the Follow Tempo functionality and have an imported loop follow your project tempo.

Figure 5.60 If you can't use the Follow Tempo feature, you can turn to manually time stretching the region so that it hits the nearest bar or locator point.

With the region/s selected (in this example, Logic can work with multiple regions to extract a more accurate calculation) select the option Edit > Tempo > Detect Tempo of Selected Region. In the accompanying dialog box, Logic will provide with a series of "best guesses," which you can select as appropriate, and then check on the Enable "Follow Tempo" or "Flex" to have the resultant calculation used to lock the loop to your project's tempo.

Figure 5.61 Use the menu options Audio > Detect Tempo so that Logic can calculate a "best guess" for the region's BPM. This can then be "flexed" accordingly.

Matching Your Project's Tempo to an Imported Loop

Now let's look at the problem another way around – where you want to change the project's tempo to match that of an imported loop. Once you've imported the region, select it, and then set the Locators to match the required bar count – like two or four bars, for example. With this established, select Edit > Tempo > Adjust Tempo using Region Length and Locators. Logic will now adjust the tempo so that it precisely fits the duration of the region, and assuming you've set the right bar count, everything should now play in time with the imported loop!

Figure 5.62 As well as getting a loop to play in time with your project, it's also possible to adjust the tempo of your project to conform to the length of your imported region.

Beat Mapping

As well as working with loops, it's also likely that you'll need to deal with performances that have little or no relation to a rigid metronome or click. In this example, you might want to "beat map" your project's timeline so that it follows the tempo fluctuations of the original performance, theoretically allowing you to synchronize additional loops, or even quantize MIDI parts, to the original recording.

Logic's Beat Mapping feature can be found as part of the Global Tracks feature (which is covered in more detail in Section 10.3). You can view the Global Track by selecting Track > Global Tracks > Show Global Track. By default, though, the Beat Mapping track lane isn't displayed, so you'll select Track > Global Tracks > Configure Global Tracks, and then ensure that the Beat Mapping option is enable.

Figure 5.63 Logic's Beat Mapping function is found as part of the Global Tracks feature. To view the Global Track, select Track > Global Tracks > Show Global Tracks.

As you'd expect with Logic, there's a variety of ways you can use the Beat Mapping feature, although in this example, we're going to explore the most common application – that of relocking a prerecorded performance and creating a new tempo grid that follows the fluctuations in the original track.

To get the process started, it helps if you find the average tempo of the performance, as well as editing the first "hit point" so that it falls at the start of your Logic project. Arguably, the best way of finding the average tempo is to use the BPM Counter plug-in, which can be found under the Metering plug-ins folder. Instantiate the BPM Counter plug-in either across your stereo bus or across one of the rhythm tracks in question. Once you've established the tempo, use the transport to set this as your project's global tempo.

Now we can turn our attention to the Beat Mapping process. Expand the Beat Mapping track and highlight the track you feel is most appropriate to work with – which will probably be something like the drum overheads. In the Beat Mapping track, you should now see the waveform of the region you've selected, and if Logic had already created transient marker information, a series of white transient marker points. If Logic hasn't got any transient marker information, press the Detect button as part of the Beat Mapping track lane.

Zoom in on the first few bars of the performance – assuming our basic tempo is correct, most (but not all) of the hits should fall on the beat. However, some of the beats will be fractionally off, possibly leading to subsequent beats also falling "off the grid." Using the Beat Mapping feature, though, we can realign

Figure 5.64
To start the Beat Mapping process, align the region/s to the start of the project and find the "average" tempo using the BPM Counter plug-in.

Figure 5.65
Step through the recording and adjust the tempo track to match the transients. First, click on the required beat in the timeline, and second, drag the "beat mapping" line to meet the transient marker in the performance.

our tempo grid to that it aligns with the hit points that we feel are important (usually on the bar, for example, or the one of the principle off-beats).

To align the beat, simply click on the beat you want to align to (on the Beat Mapping timeline) and then align this to the accompanying transient point on the Beat Mapping display. As you're aligning the beat and the transient, note how Logic draws a thin line to indicate the alignment process that is taking place. Also, the Tempo track (which is part of the Global Track) should have additional tempo points inserted into it, indicating how Logic is varying its initial tempo to stay in time with the drum performance.

Figure 5.66
Here's a finished example of Beat Mapping in action, with the associated tempo changes to keep Logic in time with the performance.

5.29   Managing Your Audio Files and Regions

In these last few sections of our audio editing chapter, we're going to review the techniques that relate to the "management" of your editing workflow – whether it's techniques to organize region data and tracks on screen so as to make your editing process clearer and easier to negotiate, or techniques to manage the "raw" audio data that are included within your project.

The Project Audio Browser versus the Project Audio Window

As we've already seen in Chapter 4, the Project Audio Browser – found to the left of the Tracks area – provides a useful overview of the audio files included within our project, alongside basic information about the files in question, such as the sample rate, bit depth, mono/stereo information, and total size. Clicking on any of the accompanying arrows next to the file opens up the list of regions associated with audio file in question.

However, in addition to the Project Audio Browser, Logic also includes a dedicated Project Audio Window (Window > Open Project Audio), which allows you to open up the Project Audio Browser separate from the main window. Although many of the functions are shared between the two types of Project Audio Browser, it's worth noting that the dedicated window includes several additional features that make it the ideal choice for more extensive file-based activates – including the ability to display full waveform views for each file/region, as well as an alternative means of setting a region's anchor point (other than using the Audio File Editor).

File Management

Using a combination of the Project Audio Browser's Edit menu and the Audio File menu, you can perform a variety of file management activities – whether it's something as simple as deleting an audio file from your hard drive or moving a batch of audio files to a new location.

One of the most useful features of the Edit menu is the option to Select Unused files. This is a great way of giving your project a spring cleaning – selecting regions and audio files that are unused in the

Figure 5.67
The dedicated Project Audio window provides an expanded set of region and audio file management options, as well as an improved graphical overview that includes waveform data.

arrangement, which can then be potentially removed from your session using the Delete key. Of course, it goes without saying that you need to approach this function with some care, although its ability to tidy up a particularly messy collection of audio files and regions can be welcome, when used correctly.

Using the option Audio File > Move File(s) is a useful way of managing particularly large audio files folders. You can choose to use "nested" folders in a way you feel is most appropriate to a project – this could involve pulling all the vocal files into one folder, for example, or grouping all the files associated with a "take" of a song (assuming you've performed a simultaneous multitrack recording across multiple inputs). The location of audio files also comes in useful with some of the viewing options that we'll see later.

Copy/Converting Files

One important task that the Project Audio Browser gets used for is the task of converting audio files between different file formats – like AIFF, WAV, or MP3 – as well as moving between different sample frequencies and bit rates. A good example of file conversion in action would be in the task of converting an original 24-bit uncompressed mix into MP3 format for use on the Internet. When you choose the conversion option (Audio File > Copy/Convert Files), Logic will present a variety of file conversion settings, as well as the option to re-reference the region in your project to the newly created files.

Figure 5.68
Use the Select Unused option to give your Project Audio Browser a spring-cleaning. Once selected, the unused regions and files can be deleted as appropriate.

Figure 5.69
The Project Audio Browser allows you to convert audio files between different bit rates, sample frequencies, and file formats, including AIFF, WAVE, and MP3.

Viewing the Project Audio Browser and Grouping Files

Using the View menu, you can configure how the Project Audio Browser displays its list of audio files and regions. One of the most useful features of this menu, though, is the ability to create a Group, allowing you to organize related regions/files so that they appear together in the Project Audio Browser. You can create a Group in a number of ways – either by selecting the files manually and then choosing View > Create Group or by using one of the automatic group options. For example, you could group files by their location on your hard drive (assuming you've used the "nested" folder previously described), their file attributes, or the current selection in the Tracks area.

Figure 5.70

Use the Project Audio Browser's Group Files feature to better organize how it displays your audio files and regions.

5.30   Using Track Stacks, Folders, and Hiding Tracks

In modern music production, it's common to have a number of "related" tracks and regions – maybe the close and ambient microphones placed around a sing drum kit recording, for example, or the multiple performances that form a "chorus" of backing vocals. As these tracks and regions amass, it can get difficult to manage the amount of material on screen, especially if you're working on a smaller-sized monitor.

One solution to a busy arrangement is to use a Track Stack, which compacts a number of tracks down to a single track lane. As well as the Tracks area being clearer, you'll also notice the mixer become slightly less cluttered. Of course, at any point in the production process, you can temporarily open the Track Stack to facilitate further editing or refinement, and then simply close the stack down once you've finished your edits. As we'll see in the Mixing chapter, the Track Stack concept can be taken even further with a Summing Stack, which provides a means for all the signals being routed to a single audio channel.

An alternative to the Tracks Stacks solution is the Logic's Folder feature, which we've already encountered to some degree with the Quick Swipe Comping feature and the idea of Take Folders. As we've seen, a Take Folder can be used as a means of packing alternative takes of the same musical line (like the lead vocal, for example) into the same folder. Once regions are packed in a Take Folder, therefore, we can use Logic's Quick Swipe Comping feature to move between the different takes.

In addition to Take Folder, though, Logic can also pack a conventional "Folder," which should be used in situations where you want to pack a number of discrete and separate regions over a number of different tracks. With the regions packed inside a folder, you can enjoy a more simplified arrangement area, with the mass of regions now represented as a single region/folder. At any point in the production, you can "peer inside" and see the contents of the folder – editing its contents accordingly, but only seeing the regions and tracks that relate to that folder. Move back out of the folder, and you'll find yourself back to the "root level" of your arrangement.

For the new users, it can be a bit difficult to discern between Tracks Stacks, Take Folders, and Folders, so here's a quick summary:

Tracks Stacks = A quick-and-easy way of merging multiple tracks lanes and regions to a single track.

Take Folder = A means of collating and editing a number of different takes of the same instrument, recorded on the same track.

Folder = A system for merging multiple regions across multiple channels into the same folder. A Folder is similar to the Track Stacks feature but more difficult to manage.

Creating a Track Stack

To create a Track Stack, select a number of track lanes in the Tracks area or mixer and then select Track > Create Track Stack. The accompanying dialogue box then asks if you want to create a Folder Stack or a Summing Stack. For mixing purposes, a Summing Stack is a good option (which we'll explore in Section 8.4, Track Stacks and the Mixer). For editing purposes, though, it's best to select a Folder Stack. Once "Stacked" you can expand the track lanes using the small arrow icon in the track list. This will also tally with the mixer, which will also flatten the tracks to a single channel strip, which can be used to quickly control the global volume level for the Track Stack.

Figure 5.71

Use the menu command Track > Create Track Stack to compact a number of track lanes into a manageable version.

Packing a Folder

To pack a folder, first select the regions you want to be include in the folder, and then choose Functions > Folder > Pack Folder from the Tracks area local menu. Logic will now create a new folder (which looks and behaves much the same as a region) as well as a folder track to play this back. Both the folder and track can be named as you see fit.

To clear some space in the arrangement, you might also want to use the menu option Track > Delete Unused. This will remove the empty tracks from the Tracks area (left when the regions were placed in the folder), although of course, they're still active "inside" the folder.

Figure 5.72
To organize complicated arrangements with multiple regions and tracks, use the menu option Functions > Folder > Pack Folder.

Despite containing a number of regions, a folder can be edited and copied in much the same way as an ordinary region – resizing it, for example, to make it shorter, or duplicating the folder so that it appears more than once in the song. As such, a folder is a very efficient way of handling multiple regions, especially where lots of edits have been made and there's a risk of losing small region fragments as you move large selections of regions around your arrangement.

Take Folders, of course, are automatically created whenever you overdub onto a track lane with an existing recording. If you want to manually pack a Take Folder so as to make use of the Quick Swipe Comping feature pick Functions > Folder > Pack Folder from the Tracks area local menu.

Viewing a Folder's Contents

You can view the contents of a folder by double-clicking on it. Logic will then present an arrangement solely of the regions (and associated tracks) that are contained within it. This will also be reflected in the Mixer, which will also display a slimmed-down set of tracks. To move back up to the "root level" of the arrangement, simply click on the back arrow in the top left-hand corner of the Tracks area.

Of course, at any point in the production process, you can choose to unpack the contents of the folder, using either the menu option Functions > Folder > Unpack Folder to New Tracks, or Functions > Folder > Unpack Folder to Existing Tracks accordingly.

Figure 5.73
Press the back arrow in the top left-hand corner of the Tracks area to move out of the folder and return back up to the root level of the Arrange area.

Hide Tracks

As an alternative to packing folders, you can also choose to, temporarily, hide a number of tracks from the current view. The Global Hide View Button can be found in the top left-hand corner of the Tracks area, and when depressed a series of buttons also inscribed "H" appear on each track beside the record-enable button. Simply choose the tracks you wish to hide, and then click on the "H" button at the top of the Arrange area. The tracks you've selected to Hide will now be out of view (and the Global Hide View button illuminated orange to indicate this), allowing you to concentrate on a simplified arrangement. When you need the tracks returned back into view, simply press the Global Hide View button once more.

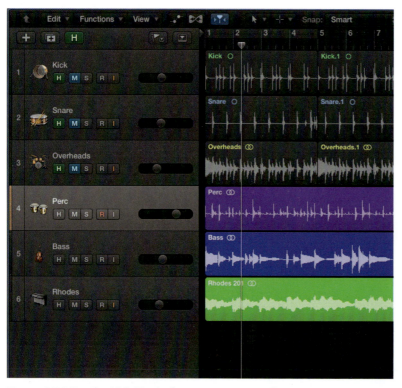

Figure 5.74 Use the Hide Tracks feature to temporarily remove one or more tracks from your view.

In this chapter

MIDI Sequencing and Instrument Plug-ins

6.1 Introduction

Besides being a tremendously competent and versatile audio sequencer, Logic has been famed for its abilities at recording and manipulating MIDI data, albeit in the virtual domain – using Logic's own virtual instruments and other third-party plug-ins – or in the now-dwindling world of hardware synthesizers and samplers. Working with MIDI in Logic offers some of the most creative and flexible musical production tools available, offering complete control of every last musical detail in your track: from the precise time and placement of notes to the exact timbre and sound parts are played with.

In this chapter, we're going to explore some of the key techniques of working with MIDI, from basic region manipulation in the Tracks area to more in-depth editing techniques using Logic's various MIDI editors: the Piano Roll, Step Editing, and the Event List. We'll also explore some of Logic's key virtual instruments – including the EXS24, the CPU-light plug-ins ES1, ES E, ES M, and ES P, alongside the Vintage Electric Piano, Vintage B3 Organ, the Vintage Clav instruments – all tools that you'll undoubtedly use on a daily basis. If you're particularly interested in the creative application of MIDI, and the techniques of sound design, be sure to read Chapter 7, which follows on from many of the concepts dealt with here.

6.2 MIDI Concepts

MIDI recording in Logic involves working with either virtual instruments embedded within the application itself (like the ES2 synthesizer or the Vintage Electric Piano, a modeled electric piano) or MIDI equipment, like hardware samplers and synthesizers, externally connected to Logic via an appropriate MIDI interface. Unlike audio recordings, a MIDI recording makes an important detachment between musical data (in other words, the raw note information that forms a musical performance) and patch data (the information required to generate the sounds). Fundamentally, what this means is that within the MIDI sequencer, we have complete and independent control over each of these elements – we can change the sound or the performance in subtle or radical ways at any point in the production process.

6.3 Creating Instrument Tracks

Creating a suitable MIDI track is done via the New Tracks command (as we saw with the creation of an audio track), although in this example, we have the choice between a Software Instrument (which would be used to access any of Logic's own virtual instruments or third-party audio units like Native Instruments' Absynth or Kontakt) and an External MIDI track, in the example of accessing legacy MIDI hardware.

Figure 6.1
Unlike an audio region, a MIDI region only contains the note data, with a virtual instrument (or MIDI synthesizer) used to control the precise properties of the sound.

The options presented for the new track types vary to that of an audio track, with less parameters to choose from. As with audio tracks, you can create multiple instances, which is handy if you know you're going to require a number of different virtual instruments in your session (16, for example, should provide an adequate starting point). If you're using third-party software instruments like Kontakt 5 that features multitimbral operation (in other words, you can load multiple sounds into the instrument), you might want to select the multitimbral option. This will create 16 different tracks, all of which are assigned to the same instrument and will allow you to address the 16 different channels, or instrument slots, within it.

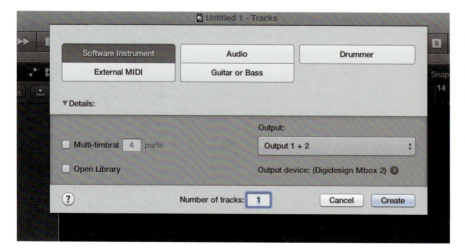

Figure 6.2
Use the New Tracks command to create a new Software Instrument track – for either Logic's own virtual instruments or third-party audio unit plug-ins like Kontakt 5.

6.4 Instantiating Virtual Instruments and the Library Feature

Assigning a virtual instrument to a corresponding Software Instrument track is done in one of two ways – either selecting from a wide palette of sounds and instruments from the Library, or selecting a specific software instrument that interests you.

Let's start by taking a look at loading in a specific software instrument – albeit one from Logic's integral instrument collection or a third-party audio unit. Possibly, the easiest way of doing this is via the Channel's Inspector using the small mixer strip that appears in the bottom left-hand corner of Logic Pro's main window. By clicking on the Instrument slot you'll open the list of Logic's integral software instruments, and at the bottom of the list, an option to browse AU instruments, which should have all your third-party audio unit instruments organized by manufacturer.

Once the instrument has been instantiated, its interface will be presented to you on the screen. On the top of the interface, you should find a drop-down list of the various presets associated with the instrument (at the moment, this will read Factory Default), as well as two arrows to sequentially move backward and forward through the list.

Figure 6.3
Instantiate an Instrument via the Instrument slot on the mixer, or the Inspector's Channel Strip.

Browsing Patches in the Library

As we've seen in previous chapter, the Library feature is the principle means of exploring Logic's vast sound content, with a range of Patches that include both the choice of instrument and accompanying

Figure 6.4
A variety of preset settings scan be accessed from the top of the instrument's interface.

Figure 6.5
Library Patches provide a complete combination of instrument and associated effects. This is a useful solution for mix-ready sounds, but the effects might begin to clutter the mix.

effects and/or MIDI plug-ins. To access the Library, use the icon in the top left-hand corner of the control bar, or use the Keyboard shortcut Y.

Arguably, what you get with the Library's Patches feature is an output that's closer to a finished sound as it might appear in the mix, although it's worth remembering that these particular settings might not be entirely appropriate to your eventual balance of sounds. Another potential downside of the extra plug-ins is the increased drain on your CPU – for example, it might be more efficient for several instruments to share a reverb plug-in rather than each instrument having its own reverb setting. Of course, there's always the option of disabling plug-ins (by Alt-clicking on them or activating their bypass control), or even removing them entirely, should you change your mind at a later point.

Plug-In Parameters versus Library Patches

If you haven't clocked it already, it's important to make the distinction between Library Patches and the selection of presets, or plug-in parameters, available from the instrument itself. As we've seen, Library

PLUG-IN FOCUS 1

Vintage Electric Piano

The Vintage Electric Piano is a virtual modeling instrument designed to replicate the classic electric piano sounds – like the Fender Rhodes and Wurlitzer – used throughout the 1960s and 1970s. Of all the vintage instrument plug-ins included in Logic, the interface is one of the easiest to comprehend, with a relatively simple set of parameters to work with. The most important of these is the Model parameter, found at the top left-hand corner of the interface, which selects a basic model from which to work. Models range from the classic Rhodes designs, like the Suitcase Mk1 and Suitcase V2, to Wurlitzers (great on rock tracks) and a Hohner Electra.

The parameters found under the Effects tab replicate the more traditional controls and features found at an electric piano player's hands. EQ and Drive, for example, govern the basic tone of the patch – keep these relatively flat if you want a clean DI'd sound or ramp them up if you need to produce the more colored tone, like the original Rhodes fed through a fender twin amp. The Tremolo control is also another feature integral to the sound of the original pianos, with controls to change the rate and intensity of the modulation. Try experimenting with the stereo phase – the 180° setting produces a pleasing stereo autopan, while setting the modulation progressively nearer 0° produces an increasingly mono Tremolo. The Chorus and Phaser are also important Rhodes-orientated effects used to great effect in the late 1970s on tracks like 10cc's "I'm Not In Love."

Experimenting with the model parameters found under the Details tab allows you to twist certain "mechanical" properties of the original instruments – variations that might have occurred with the condition of the instrument, for example, or in the way it was manufactured. For example, increasing the Tine Bell volume makes the sound closer to the bright "Dyno Rhodes" modifications performed on Fender Rhodes pianos in the early 1980s.

(Continued)

Figure 6.6 The Vintage Electric Piano remains one of the silkiest electric piano emulations available, with a surprisingly low CPU drain.

Figure 6.7
The Library provides a more elegant and seamless way of stepping through an instrument's plug-in parameters.

Patches contain additional effects, and/or MIDI plug-ins, to make them a "complete" solution, whereas plug-in settings are just a snapshot of the instrument's current state. You could, for example, load a Patch, a scroll through different plug-in parameters while the effects remain intact, or load a completely different Patch and have the effects changed with it.

Once you've made this distinction, it's easy to use the Library to navigate both Patches and plug-in parameters using the small blue triangle that floats to the side of the Inspector's channel strip. If the triangle is at the top of the channel strip, to the side of the Setting box, then the Library will be used to navigate Patches. If the triangle is moved down the channel strip, though, toward the instrument selection, then you'll be browsing the instrument's various parameter settings.

Figure 6.8 The position of the small blue triangle indicates whether you load full, Library patches with effects, or simply move between different plug-in parameter settings.

Changing the Track Parameters

Track Parameters are a global set of parameters associated with a particular instrument. The Track Parameters can be found via the Inspector, with the current selected track having its parameters displayed toward the middle of the Inspector – remember to open the box (using the small arrow) if it is minimized. Although the Track Parameters include some useful features, most users tend to leave these settings in their default positions, unless they're working with external MIDI sources (see Section 6.5 for more information on this).

Figure 6.9
The Track Parameters box allows you to control a number of key parameters in relation to the instrument instantiated into a particular track. On the whole, this can be left in its default state.

Changing an Instruments Icon

Icons allow you to easily distinguish between different tracks in your Logic project. By default, any of Logic's instruments will establish a corresponding icon to distinguish between the different models (EXS24, ES2, and so on), although you might want to use one of Logic's other icons to define a track's musical properties (strings, bass, guitar, drums, and so on). To change the icon, simply click on the icon setting in the Track Parameters box.

Figure 6.10
Using icons will allow you to better distinguish between different tracks and their various musical roles.

Transposition and Velocity

Use the transposition and velocity parameters to quickly change the way an instrument responds to the performance data it's presented with. For example, if you know you want the instrument to take a prominent, aggressive position in the mix, it might be appropriate to use an additive value (like 120) for the velocity setting. All performance data sent to this instrument – whether played live from MIDI keyboard or from any track in the Tracks area assigned to this instrument – will be modified accordingly. If you want more control as to when these modifications occur, consider applying them as MIDI edit – either by modifying region parameters or by using one of the MIDI editor windows, both of which we'll explore in more detail later on in this chapter.

Key Limit

This feature is useful where you're orchestrating for acoustic instruments with a known limit in their playable range. For example, a first violin could have its lower range set to G2 to remind you not to play or record any MIDI data below this point. Although some of the sample instruments included with the EXS24, or as third-party data, might account for the natural instrument's playable range, not all do. A little forethought here, therefore, could save hours of extra editing and arrangement work later on.

6.5 Working with External MIDI Instruments

First off, if you haven't got any external MIDI equipment, you can skip this section and carry onto the next! However, if you're lucky enough to have some external MIDI sound sources, like (proper) vintage synthesizers and samplers, then you'll need to spend a few minutes adjusting a few settings to get things working correctly. It could also be the case that you're also using the MIDI protocol to communicate with other computers in your studio, even using a network connection to connect to an additional machine running a software sampler in standalone mode, in which case you'll also need to follow these basic configuration steps.

Because we're working with an external MIDI device, we'll need to ensure that our track is being directed to the right physical MIDI output on our corresponding MIDI or audio interface. Start by selecting the External MIDI track in question, opening the Inspector, and looking at the Instrument Parameters. You'll need to establish a Port setting to correspond to the port your MIDI device is plugged into, as well as setting a corresponding MIDI channel. If the instrument is multitimbral – in other words, a device that allows a different sound to be placed on each of the 16 available MIDI channels – you might want to run multiple External MIDI tracks, each set to a different MIDI channel. To do this, go to the Track menu and select Other > New with Next Channel. You can also Ctrl-click at any point on the track name to reassign the object to another MIDI channel accordingly.

Figure 6.11
Change an External MIDI Instrument's Port setting (as part of the Track Parameters) to assign it to a physical connection on your MIDI/audio interface.

PLUG-IN FOCUS 2

Vintage B3 Organ

The Vintage B3 Organ is Logic's own virtual recreation of the classic Hammond organ. The original instrument used a unique tonewheel sound generation system, controlled by a series of drawbars, to create a surprisingly rich and varied set of organ sounds. The package was then made complete by the addition of a rotating speaker cabinet – otherwise known as the Leslie – providing a distinctive swirling movement to any sound that passed through it.

Like the original B3, the Vintage B3 Organ supports two manuals (or keyboards) and a set of bass pedals, each with its own set of drawbars. To access these, you'll need to set the MIDI channel in the instrument's parameter box to All – this way the EVB3 will be able to receive MIDI data on all channels, with Channel 1 being directed to the upper manual, Channel 2 to the lower manual, and Channel 3 to the bass pedals. Alternatively, change the mapping via the Split tab, where you can switch between the aforementioned Multi Mode (where each manual is accessed via a different MIDI channel) and instead use the Split mode to have the various manuals accessible from different octaves of your controller keyboard.

Shaping the tone of the Vintage B3 Organ is achieved by sliding the relative position of each manual's drawbars on the Main screen. Essentially, the drawbars represent a different harmonic from the harmonic series – try using a simple combination of just a few of the lower drawbars for a mellow, pure tone, or "pull out all the stops" for a fuller timbre that is rich in harmonics. The colors also provide some indication to their respective roles: the white drawbars being even harmonics,

(Continued)

black drawbars containing the odd harmonics, and the brown drawbars representing the sub (an octave below the fundamental) and third harmonics.

Switching between the different Leslie settings found under the Rotor Cabinet screen (Chorale, Brake, and Tremolo) forms a distinctive part of the movement and progression in a Hammond performance. The Chorale creates a slow vibrato, the Tremolo a fast-moving vibrato, whereas the Brake stops the vibrato effect. In addition to being able to switch the settings from the interface (these could, for example, be recorded into MIDI), you can also use the mod wheel to move through the three settings.

Figure 6.12 The Vintage B3 Organ is Logic Pro's take on the classic Hammond organ, with two manuals and a set of bass pedals, and of course, the authentic drawbar sound creation system.

How Do I Monitor MIDI Sound Sources?

Although you may have routed MIDI out of Logic to the device in question, there's still no guarantee that you'll be able to monitor it correctly. Unlike Audio Instruments, of course, an external MIDI instrument won't have a corresponding mixer channel to monitor its output (although you do have a fader to remotely control the level and pan settings via MIDI). One solution is to run your main outputs from Logic into a physical mixer (Channels 1 and 2), with your synthesizer, sampler, or drum machine connected into any remaining spare channel. With this setup, however, you won't be able to use any of Logic's own effects on the synthesizer's output, or indeed, use the bounce to disk mixdown functionality to create a complete version of the mix.

A more appropriate solution is to plug the outputs of your MIDI hardware sound sources into spare inputs on your audio interface. Next, create a number of spare audio tracks (one for each device connected to your audio interface), with each of the audio tracks set to Input Monitor. Return to your External MIDI track, press a few keys on your controller keyboard, and you should now hear the sound source being

Figure 6.13
To monitor a MIDI synthesizer, you'll need to connect it to your audio interface's audio inputs, create a corresponding audio track, and place this into Input Monitor mode. This will also eventually allow you "print" the synthesizer's output as an audio file.

routed back into Logic's main mix. Note that you might need to adjust the input gain on your audio interface to account for the levels being sent out from your sound source, and that the monitoring will, of course, be subject to any latency present in your system.

As an alternative to creating a new audio track, try using the "Use External Instrument plug-in" option when you're creating a new MIDI track. The newly created track will look much like a traditional virtual instrument, although you'll notice that the External Instrument plug-in has been instantiated in the Instrument slot. Opening up the parameters setting for that Instrument, you should find options to specify both the MIDI destination (used to direct the MIDI performance data out from Logic) and the Input selection, used to return the audio signal back into Logic's mixer.

The significant advantage of your MIDI sound module being routed back into Logic (using either the manual routing technique or the External Instrument plug-in) is that you have complete control over how it appears in the mix, even with the possibility of adding plug-ins, like equalizers, compression, and reverb. If you've picked the manually routing option and you're completely happy with the performance and sound, you can always quickly flick from Input Monitor to Record and have the part permanently stored as an audio file, which might offer a better long-term solution, as well as being a convenient means of negating any latency issues.

What If I Have More Than One External MIDI Device?

Adding further MIDI devices beyond the basic GM Device included in the basic Empty Project file goes beyond the scope of the main Tracks area interface. Instead, you'll need to create new instrument objects as part of the environment, which is something we'll cover in more detail in Chapter 11. If you have a particularly complicated MIDI setup, you might want to consider building your own Template file so that all your preexisting MIDI devices are automatically set up with each new project. Again, this is something we'll specifically look at in Chapter 11.

My MIDI Tracks Are Playing Back Slightly Late, How Do I Rectify This?

As we mentioned before, it is feasible that the monitoring of External MIDI tracks via your MIDI hardware could be causing some minor latency issues, with the MIDI tracks appearing to play slightly behind the beat. To rectify this, you might want to consider applying a negative delay via the Instrument Parameters so as to send the MIDI data out slightly ahead of time and therefore negate any of the latency involved. Finding the right settings (which will probably be in the region of about – 20 or so, depending on the latency in your system) could be as simple as adjusting the delay parameter by ear. Alternatively record in a bar or so of MIDI information as an audio file, and then adjust the instrument's negative delay until the "live" MIDI data and recorded MIDI data playback exactly in sync.

One important factor to note here is that once a MIDI track has been recorded as audio, the monitoring latency will be removed. It is important, therefore, that if you intend to record an external MIDI part into Logic, you ensure that the delay parameter has been reset to its zero point.

To throw yet another wrench in the works, it should also be pointed out that Logic's plug-in delay compensation (found under Preferences > Audio > General), where used with latency – including DSP – acceleration systems like Universal Audio's UAD-2, can also create additional MIDI playback latencies that

Figure 6.14
Adjust an Instrument's delay settings to account for any monitoring latency.

will vary given the amount of plug-ins, and where they are placed in Logic's mixer. The best advice here is to avoid applying these types of plug-ins until the mix or use the low-latency monitoring mode to temporarily bypass the offending plug-ins while you are monitoring External MIDI instruments.

6.6   Making a MIDI Recording

Making a MIDI recording follows many of the same procedures explored in Chapter 4, although you might want to note a few dissimilarities.

Where's the Data?

The MIDI data created with a recording stays within the region, rather than referencing a separate audio file, as is the case with an audio region. MIDI data takes up an incredibly small amount of memory, with a whole project's MIDI resources utilizing less than about 1 MB of RAM in total.

Recording in Cycle Mode and Overdubs

Unlike an audio recording, which will create a Take Folder for parts recorded in Cycle mode, a MIDI recording will simply layer addition passes into the same region. The assumption with Cycle mode is that you use the repeating playback to record each part a layer at a time (like a kick, followed by the snare), without stopping the transport at any point. If you want to record different takes of a keyboard solo, for example, consider creating a new track assigned to the same instrument (Track > Other > New Track with Same Channel), with the previous take muted. Recording directly over the first take (as you would with an audio overdub) will simply create regions placed on top of one another, all of which will play at the same time. Multiple stacked MIDI regions – unlike Take Folders – can get difficult to manage, so try to keep on top of any layered recordings using the editing and arrangement techniques described in the following parts of this chapter.

Figure 6.15
Multiple MIDI recordings will simply create "layered" takes, which can get confusing on a single MIDI track. Consider creating duplicate Instrument tracks to better distinguish between your various takes.

Figure 6.16
In this example, a duplicate track has been used to record the second take of a MIDI performance, rather than layering it onto the original MIDI region.

PLUG-IN FOCUS 3
Vintage Clav

The Vintage Clav is Logic's virtual recreation of Hohner's "funkalicious" Clavinet. The original instrument worked like an electrified harpsichord, with a number of reeds plucking a series of strings, with the resultant vibrations being picked up by a collection of guitar-like pickups mounted in the unit.

As with the Vintage Electric Piano, the Vintage Clav divides its interface up into Main tab; with a series of conventional controls that you'd find on the original instrument, an Effects tab; with various signal processing option, and a Details tab; for fine tuning the modeling parameters to adjust every last detail of the Vintage Clav's output.

The conventional controls focus on the rocker switches toward the bottom of the interface. These move between basic, preset filter positions (brilliant, treble, medium, and soft) and different pickup combinations. For example, by alternating the various filter rockers, you can move between a soft, almost muted tone, to a full and raspy clavinet pluck. The pickup switches produce more subtle transformation in the sound and are worth experimenting with to extract just the right flavor from the instrument. It's also worth noting that you can physically move the pickup – using the Pickup Position graphic – which is an interesting alternative means of changing the tone of the instrument.

In keeping with the approach of many keyboard players in the 1960s and 1970s, the Vintage Clav includes its own Effect section covering Distortion, modulation effects (Phaser, Flanger, and Chorus), and the infamous wah-wah. Although the wah-wah can be set to track the sound's envelope (for an autowah effect), the best solution is to map it to something like the modulation wheel (found under the Details tab), or a spare MIDI controller, to keep it as hands-on as possible.

Figure 6.17 Logic Pro's Vintage Clav is a faithful recreation of Hohner's "funkalicious" Clavinet – complete with the original controls and the number of modeling options.

6.7 Editing and Arranging MIDI Regions

Again, the process of editing and arranging MIDI regions in the Tracks area follows many of the same conventions as audio editing, allowing you to expand and develop your music arrangement over time. As we saw in the audio editing chapter, regions can be resized, cut, moved, and duplicated using the same tools and processes involved in audio editing. However, as with the process of recording, it's worth pointing out a few differences with respect to working with MIDI regions.

Merging MIDI Regions

Using the glue tool allows you to merge a number of MIDI regions into a single contiguous region, although unlike audio data, this doesn't result in the creation of a newly rendered audio file. Merging MIDI regions is a great way of combining any stacked MIDI regions (where regions have been stacked on top of one another), allowing you to see all the MIDI data in a unified way.

Working without Fades and Crossfades

As you'd expect, a MIDI region doesn't require any crossfades to be placed at edit points; therefore, it is impossible to apply fade-based processing to a MIDI region. To achieve the same volume contouring, you'll need to turn to either Automation (for more information see Section 8.10 in Chapter 8) or the MIDI Draw feature (covered in Section 6.17).

Timestretching MIDI Regions

Although it is possible to perform a certain level of time adjustment with audio regions, it is MIDI regions that offer some of the greatest flexibilities with respect to time-based properties. By Alt-dragging the end point of a MIDI region, for example, you can quickly time compress or expand the data that are contained within it. This can be an interesting way of exploring double-time or half-time treatments by stretching the region by 200 or 50% of its original size. Of course, as this is simply the playback of MIDI data you're affecting, there is no loss to the sound quality of playback, which would easily become apparent with the same level of stretch applied to an audio file.

It should also go without saying that a MIDI region follows and responds to the project's tempo, without needing to activate the Follow Tempo feature in the Region Parameters box. This also makes it feasible to record a MIDI region at more comfortable playing speed and then increase Logic Pro's tempo to playback the region at the appropriate speed. The only issue to be aware of here, though, is accompanying audio tracks, which won't necessarily follow the same tempo change. In this case, simply monitor from the virtual instrument or MIDI tracks accordingly.

Figure 6.18
Using the Alt key as you resize a MIDI region will allow you to time compress or expand its MIDI data accordingly.

6.8   Region Parameters: Quantizing and Beyond

Once you've recorded your first few bars of MIDI data, you can now start to explore some quick-and-easy modifications directly from Logic Pro's main window. One of Logic Pro's key concepts is that a large number of MIDI modifications can be performed directly from the Tracks area, without ever having to go anywhere near a conventional MIDI editor (like the Piano Roll or List editor windows). To perform these modifications, simply select one or more regions and turn your attention to the Region Parameters box as part of the Inspector.

The main parameters accessible from the Region Parameters box include the region's name, quantizing, loop, transposition, and velocity settings. What the Region Parameters allow you to do is to quickly define some important overriding concepts of how the region – or group of regions – is played back. None of the original "recorded information" is modified in this action – these parameters are simply "playback" criteria. For example, you could select the entirety of your MIDI composition and adjust the transpose option to find a key that best suits your vocalist's range. Should you decide to return your composition to its original pitch, simply select the regions and reset the transposition accordingly.

Figure 6.19
The Region
Parameters –
available via the
Inspector – allow
you to quickly
define key
properties of
your MIDI regions.

Quantize

Arguably, the most important function of the Region Parameters is the Quantize feature, allowing you to quickly tighten the rhythmic qualities of a given performance. Clicking toward the right-hand side of the Region Parameters box allows you to open up a list of the Logic's available quantize options, ranging from various 1/8, 1/8 Triplet (1/12), and 1/16 note settings, to Swing quantizing, and combined 16/24 options. On the whole, most conventional musical applications will adopt a simple 1/16th note quantizing so that any stray notes are pulled to a rigid sixteenth (or semiquaver, to give it its musical name) grid. A Swing quantize, on the other hand, shifts every other sixteenth, producing a distinctive shuffle effect commonly used in Hip-Hop or R n B productions. In Swing quantize, the various letters (1/16 Swing A, 1/16 Swing B, 1/16 Swing C, and so on) denote the relative amount of shuffle, with 1/16 Swing A, for example, delivering a lighter shuffle than 1/16 Swing F.

Figure 6.20
Here's a simple MIDI sequence before and after quantizing. Note how the timing of the quantized region (top) conforms exactly to the project's tempo grid.

Q-Swing

The Q-Swing offers a finer control over the swing setting than the 1/16 Swing A, 1/16 Swing B, and 1/16 Swing C settings found in the standard drop-down Quantize menu. At 50%, the feel is straight, while settings either way (60% or 40%) imparting different qualities of shuffle. If you're particularly into the groove of a piece of music, then the precise control offered here will be important in creating the right feel between parts.

Loop, Transpose, and Velocity

Loop – as with the same function in an audio region's Region Parameters – can set an indefinite loop of the MIDI region until either the end of the project or another sequence object is encountered. Transpose, of course, deals with the relative pitch of the MIDI sequence, allowing you to transpose a region, or group of regions, up or down by a selected amount of semitones. Finally, Velocity allows you to control the relative verocity that a part is played at – setting a number in the velocity parameter, for example, will either add or subtract a value to the current velocity levels contained within the part.

PLUG-IN FOCUS 4
ES E, ES P, and ES M
Although it's easy to get carried away with many of Logic's deeper and more complex software instruments, there's still a lot of use to be extracted from some of its simpler synthesizers. One of the best qualities of these instruments is their simplified user interfaces, often with a bias toward specific sound types, like pads in the case of ES E, or leads and basses in the case of ES M.

(Continued)

Arguably, the simplest synthesizer in Logic is the ES E, which is dedicated to simple pads sounds, all from a single oscillator. Try experimenting with some of the key parameters (in other words, the ones with the bigger knobs), exploring the different waves on offer, and tools like cutoff and resonance. The three modes of Chorus (Chorus I, Chorus II, and Ensemble) are vitally important in creating warmth to the pad sound, with each setting offering a slightly different flavor.

ES M is a virtual instrument modeled on the simple synthesis capabilities of early monophonic synthesizers, like the Roland TB-303. This is a great source for "acidic" basslines, full of distortion and resonance, as well a simple lead synthesizer parts. The Glide parameter is an interesting feature – and a quintessential part of the aforementioned TB-303 bass sound – allowing the ES M to gliss from one note to the next whenever the first note is held over.

The ES P offers the most complex set of sound possibilities of the three synths and can create a range of different synth sounds. One of the most distinctive features is the waveform mixer, which allows you to blend a number of different waveforms to create a starting point for the rest of the synth. The first three sliders govern basic waveform types (triangle, sawtooth, and square), while two of the final three sliders add square waves an octave below and two octaves below that of the fundamental pitch. Further sound-shaping features include envelope generators, Chorus, and an overdrive section.

Figure 6.21 Logic Pro's CPU-light collection of synths can be a great place to turn to for some simple keyboard sounds: like 1-oscillator synth basses or sawtooth pads. Note how each synth is dedicated to a certain range of sounds, making the interface uncluttered and easy-to-use.

6.9 The MIDI Thru Function

If no region is selected in the Tracks area, you should notice that the Region Parameters box displays MIDI Thru rather than an appropriate region name. As innocuous as the MIDI Thru function may seem at first, it is, in fact, a particularly powerful feature of sequencing in Logic Pro. MIDI Thru's most immediate and noticeable benefit is as "live" modifier of MIDI data coming into Logic Pro. For example, rather than playing in a difficult key, like C# major, you could choose to play in C major and use the transposition parameter to shift everything that you play up a semitone. Equally, the velocity parameter can be a useful way of increasing or decreasing the relative loudness as you play.

Another important function of the MIDI Thru option should also become evident as you make recordings into Logic Pro. In effect, the MIDI Thru function sets a default Region Parameter setting for everything

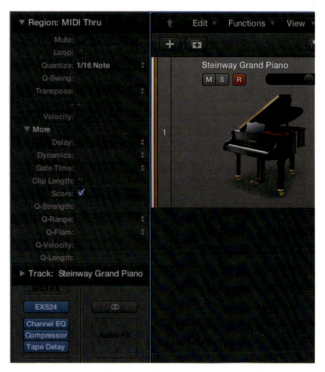

Figure 6.22 Using the MIDI Thru option as part of the Region Parameters can preset certain playback criteria – like quantizing or transposition, for example.

you record into Logic Pro. Going back to the transposition example, you should note that a part recorded in with MIDI Thru set to +11 transposition will have its Region Parameter set accordingly (this also means you can go back to the data in its original form). This MIDI Thru functionality is particularly powerful with a setting like Quantize, where once set, all parts will be automatically quantized on-the-fly, immediately after they've been recorded.

6.10 Advanced Quantization Options

In addition to the basic quantizing options we've covered so far, Logic Pro also includes an expanded set of quantizing features that facilitate a more refined and delicate approach to quantizing – from the ability to "softly" quantize a MIDI region, for example, through to creative possibilities like Flam quantizing. Although slightly more perplexing for novice users, Logic Pro's advanced quantizing options are essential if you want to retain some of the feel of the original performance and avoid the overtly "robotic" sound of over-quantized music.

You can view the advanced quantizing options by clicking on the small "More" arrow toward the bottom of the Region Parameters box. Once activated, you should find the Region Parameters updated to include a further set of options. In addition to the advanced quantizing options, you'll also see an expanded set of Regions parameters covering additional areas like Delay, Dynamics, and Gate Time.

Figure 6.23
The extended Region Parameters offer more precise control over the groove and strength of quantize applied. Use these where you need to preserve important qualities of the original performance.

Q-Strength

Arguably, one of the most important options in the extended Region Parameters is the Q-Strength parameter. As the name suggests, this directly modifies the "gravitational" strength of quantizing so that a 100% setting, for example, pulls the notes completely to their grid position, while a setting of 50% only brings the notes halfway. This is a great way of achieving a soft quantizing – improving the overall feel of the performance, without turning it into a metronomic sequence.

Q-Flam

Q-Flam avoids the sometimes-unnecessary side effect of a quantized chord having its notes aggressively bunched together. By using the Flam setting, you can "fan-out" the notes, with positive settings producing an upward movement and negative settings producing a downward movement. Although this won't be applicable to every instrument you work with, the Flam option is a feature that works really well with harpsichord or acoustic guitars samples.

Q-Range

One of the more difficult tools in the quantize tools set to use is the Q-Range control. Q-Range is another form of a soft quantizing tool, although in this case, the notion is to ignore notes closer to or on the beat, and instead direct the quantizing engine toward notes that stray beyond a given point. On a setting of zero, for example, all notes will be corrected to the current quantize settings. By decreasing the range into its negative setting, only notes that fall outside the range will be quantized, effectively creating an increasingly soft form of quantizing.

Dynamics

The dynamics parameter is interesting, in that it works like a MIDI compressor or expander – reducing or expanding the range of velocity levels that occur in the region. Indeed, couple these functions with an expressive virtual instrument, and you can have a surprising impact on how the instrument sits in your mix.

Gate Time

Gate time deals with the duration of notes – changing an elongated series of notes, for example, into a shorter staccato effect. You can achieve some interesting effects here on repetitive sequencer lines, especially where all notes are of fixed duration. Try experimenting with the different gate times and see how this affects the finished output.

6.11   Normalizing Sequence Parameters

Should you reach a point where you're completely happy with your region's parameters, you might want to consider making the "adaptive" edits permanent – a process known as Normalizing. Although you'll lose the possibility of going back to the original performance, a normalized region might be easier to handle in the long run, especially if you manage to undo your perfected edits by mistake.

The appropriate various normalize functions can be found under the MIDI menu option when you ctrl + click on a MIDI region, with various options to Normalize Region Parameters or destructively apply any Quantize settings you've created.

Figure 6.24
Use the Normalize feature to permanently "fix" your Region Parameters like quantizing.

6.12   MIDI Editing in Logic

Although there's a surprising amount that can be achieved using just the Region Parameters alone, eventually you hit a point where you need to explore a more detailed set of editing options.

Editors can be opened in a number of ways. For example, double-clicking on a region will open the default assigned editor (Logic Pro X > Preferences > General, under the Editing tab), which is initially set to the Piano Roll editor. Alternatively, once one or more regions are selected in the Tracks area, you can also use Editors button on the control bar and the various Piano Roll, Score, or Step Editor tabs at the top of the editing pane to open the respective editors. Finally, using the Window menu, you can also open separate editing windows away from Logic Pro's main window, which is a useful solution if you're using multiple displays.

Delving deeper still, it's also possible to view and edit your MIDI data in the form of a list of Event data – in other words, the precise note on, note off, and controller movements that your sequence is composed of. Although this might seem confusing at first, it does offer a precise and informative view of your MIDI data (much the same way Logic sees it), allowing you to position MIDI notes on a tick-by-tick basis, or set controllers to exact parameter levels.

Ultimately, good MIDI editing practice in Logic is about matching the right MIDI editor with the precise type of modification you want to achieve. For example, if your edits are generally musical, look toward the Piano Roll or Score Editors. If your edits involve a more intricate manipulation of control data, then either the Step Editor or Event List editor might be more appropriate. Not only will the right editor allow you to work quicker, but also might allow you to discover some interesting creative possibilities along the way.

6.13   The Piano Roll

The Piano Roll editor provides one of the most intuitive and informative ways of editing MIDI data in Logic and is often the first choice for the majority of MIDI editing tasks. The editor can display both MIDI note data in the top half of the interface, with the added option of viewing controller data, like pitch bend and modulation toward the bottom of the editor. Notes are displayed in much the same way as regions in the Tracks area – with the horizontal axis indicating their timing and length, while the vertical axis indicates their relative pitch. In addition to the pitch, timing, and length of the notes, you can also see their velocity, displayed as a series of colors.

Figure 6.25
Here's a typical MIDI region being edited in the Piano Roll editor, with note data toward the top of the interface and controller data toward the bottom.

PLUG-IN FOCUS 5

EXS24 Main Interface

As a completely integrated solution for using samples, Logic's EXS24 is one of its most powerful music production tools. Compared to other software samplers, like Native Instrument's Kontakt 5, the EXS24 appears a much more approachable tool, although this is largely because

(Continued)

it makes a distinction between the front-end controls (which we're looking at here) and the behind-the-scene mapping of samples (which we'll explore in more detail in Chapter 7).

The interface is divided into a number of key sections. Most important is the Sample Instrument menu, with the current selected instrument displayed in the top part of the ESX24's interface. Clicking on this will call up the list of EXS24 Sample Instrument stored on your Mac, as well as Sample Instrument that you might have specifically created for the project in question. You can also browse through, instrument by instrument, using the small + and − buttons on either side of the Sampler Instrument menu.

Once a sound has been loaded in, you can then start to experiment with the various parameters settings on the interface. These include basic instrument tools – like volume, tuning, total available voices, and so on – as well as a number of synthesis-orientated features, like a multimode filter (just below the Sampler Instrument menu), three LFOs, two envelope generators, and a modulation matrix. In theory, taking the basic mapping of the samples, you can use the range of synthesis tools to transform the basic qualities of your sound, making its timbre darker, for example, or using the envelope to change the attack characteristics. We'll be taking a more detailed look at these possibilities in Chapter 7.

Unlike certain other third-party software samplers, the EXS24 isn't designed to work with multiple instruments in the same instance. Being directly part of Logic's code, the EXS24 is extremely CPU-efficient, with the intention that you run multiple instances for each sample instrument you require in the mix.

Figure 6.26 Unlike other third-party samplers, EXS24 is specifically designed so that you load multiple instances – one EXS24 for each sound, or instrument, you want to appear in the mix.

6.14   Typical Editing Scenarios in the Piano Roll

If you're used to editing regions in the Tracks area, you should find that many of the techniques carry directly over to the editing in the Piano Roll editor. As with the Tracks area, this editing process is somewhat tied up with a series of tools, including familiar objects (like the Pointer tool, Pencil tool, and so on) alongside two tools that are unique to the Piano Roll – the Quantize tool and the Velocity tool.

To understand how the various tools work and relate to each other, let's look at a variety of editing scenarios and how you might combine the various tools and techniques in the Piano Roll editor to rectify this situation.

Correcting Pitch

Using the pointer tool, you can move a note vertically up or down to change its relative pitch. By adding in a few keyboard modifiers, you can also perform some quick modifications without needing to drag objects about. For example, press Alt + either the up or down arrows and you can move a note or selection of notes up or down in semitone steps. Add in shift (Alt + Shift + Up/Down) and you can move the same notes in octaves.

Figure 6.27
Use Alt + Shift and either of the up or down keys to move selected notes in octaves.

Correcting Timing by Hand

Although quantizing can offer a quick-and-easy route to correcting the timing problems of a particular MIDI sequence, sometimes it can be important to have a more precise control over the placement of notes in your sequence. By leaving the region unquantized, you can use the pointer tool to manually position single notes, or groups of notes, by hand.

As with audio editing, it's important to note the use and operation of the Snap feature (found in the top right-hand corner of the Piano Roll) and how this interacts with the movement of the notes on the Piano Roll's grid. In most cases, this is set to its Smart setting, allowing you to place the notes anywhere along the timeline, although with a definable "pull" to the grid points. If you specifically want the notes to always snap to the grid, you can change this to beat or division, for example, depending on how fine you want the resultant grid. For truly precise movements, though, it's worth adding in a few keyboard modifiers. Shift + Ctrl will switch off any grid setting – irrespective of the current Snap setting – allowing you to precisely move the note in question. Alternatively, use the Nudge feature (Alt + the left or right arrow) to move the note tick-by-tick using the keyboard.

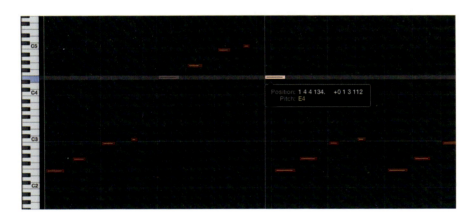

Figure 6.28
Holding Shift + Ctrl as you move a note allows you to negate any preexisting Snap settings and place each note to a precise tick-based position.

If you want to get creative, it's also worth experimenting with the timing of phrases by shifting them about in various divisions. For example, try taking an existing phrase and moving this by 1/16th, 2/16ths, or 3/16ths. Notice how the feel of the phrase has changed with the notes falling in a different placement. You can also use this to good effect by starting a track with a phrase that doesn't fall on the first downbeat of the bar, and then, as the drums come in, the listener experiences mental "shift" as they're forced to relocate the bar.

As you'd expect, there's also a sophisticated range of quantizing tools and options available from within the Piano Roll editor. This will be discussed in Section 6.15.

Duplicating a Phrase

By adding in the Alt modifier as you drag your selection, you can create a copy of the note(s) in question. This is an effective way of duplicating musical material within the region.

Correcting Length Issues

The exact length of a note can be as important to the groove of a part as the note's start point. The specific tool assigned to this task is the Finger tool, although you can achieve a similar result using the Pencil tool and Pointer tool by clicking toward the end of a note's duration. Try selecting one or more notes and reducing their lengths to create a sharp, staccato-like playing style. Alternatively, try lengthening the notes to produce a smoother, more legato-like effect. Mixing and matching these different length settings can be surprisingly effective on basslines.

Figure 6.29
Use the Finger tool to adjust the length of one or more selected notes – moving between a staccato and legato playing style.

Changing Velocity

Velocity – or in other words, how hard you play a note – has a big part to play in the expression in your music. Ideally, the best types of music incorporate a range of different velocities – both from note to note and over the whole "emotional arc" of the piece. The Velocity tool in the Piano Roll editor allows you to change the velocity of any note or group of notes you click on. Simply click on the note and move the mouse up or down accordingly to increase or decrease the current velocity accordingly. Where a group of notes is selected, the velocity will keep their respective balance (in other words, some notes louder than other).

One aspect of editing where velocity really comes into play is drum programming. As you're putting together a pattern, be sure to experiment with different accents on the various main beats, as well as adding discrete ghost notes using lower velocity settings.

Although the Velocity tool provides one solution to changing a note's loudness, it's also possible to use the MIDI draw to also achieve the same task. Have a look at Section 6.16 for more information.

Figure 6.30
Here's an example of a programmed drum loop making extensive use of the velocity tool to define the accent and grove of the loop.

Creating New Notes

Use the Pencil tool as a means of creating new notes in your region. As we saw in the section on changing length issues, the Pencil tool can also be used to define the note length by clicking toward the end of the block. One other trick to look out for is how the Pencil tool defaults to creating a note based on the last edited note event. For example, if you create a note and modify to be 1/64th long with a low velocity, all subsequent notes created with the Pencil tool will conform to that standard. This can be a useful way of setting a default note type for all subsequent newly created notes to follow.

KNOWLEDGEBASE 1
EXS24 Virtual Memory and Disk Streaming

In an attempt to make a realistic recreation of acoustic instruments, many third-party sample libraries can make quite exhaustive demands on your RAM. For example, some Piano instruments can comprise multiple velocity layers, with a separate set of samples for each key on an 88-note piano keyboard. It's not uncommon, therefore, for these instruments to have RAM demands between 1 and 2 GB – something that many users' systems will fail to cope with.

If you do run out of available RAM (the sampler will tell you that it has been exceeded and will fail to achieve a complete load), you might need to consider using the EXS24's Virtual Memory

(Continued)

feature. Virtual Memory works by only loading a small amount of the start of each sample, with the remainder streamed from your hard drive as and when required. By using Virtual Memory, you'll be able to load a far greater number of these large instruments into a project and have them played back successfully.

To access the Virtual Memory feature, open up the EXS24 Options menu and select Virtual Memory from the list. A dialog box will allow you to set and optimize settings, balancing off the hard drive's speed with the amount of disk activity (namely, audio track streaming and recording) also happening at the same time. Obviously, if you're already running a lot of audio tracks, this might begin to put your hard drive under an increasing amount of strain, whereas a low amount of disk activity will ensure optimal performance with a large amount of sample voices being playable at the same time. You might also want to consider using an external Fire 800 drive, or an additional Serial Advanced Technology Attachment (SATA) drive (internal or external), to keep certain types of data accessible quickly and easily.

In most cases, the usual "Core Audio Overload" will indicate the system being pushed too hard, although you might also want to listen out for longer notes tails, which can get snatched off by excessive disk activity. Also check out the Disk I/O Traffic and Not Reading from Disk in Time performance indicators in the Virtual Memory Dialog box. These will also indicate the effective, or problematic, use of your system resources.

Figure 6.31
With Virtual Memory engaged, you'll be able to load a greater number of RAM-hungry software instruments. Be aware, though, of potential knock-on effects on your system's performance.

Deleting/Muting Notes

To remove a note from a region, use the delete tool or simply select the required notes and hit delete on your keyboard. However, a more strategic solution might be to use the mute tool. By muting the note event, it will cease to be played back from the region, but at any point, you can return to the region and unmute the event to hear the original note as it was once played.

One other useful application of the muted notes feature is as an editing guide. For example, say you're trying to harmonize or augment an existing sequence using a different virtual instrument. By copying the region and muting the original notes, you can visualize what the other part is playing, while at the same time constructing or editing a new sequence to play over the top.

Figure 6.32
Muting notes is a useful alternative to delete them. In this example, a muted line has been used as a guide to create a harmonized part.

6.15   Quantizing Inside the Piano Roll Editor

Despite the Inspector being the first port of call for quantizing tasks in Logic, it's also interesting to note that another layer of quantizing exists within the Piano Roll editor itself. As flexible as this is, this two-tier system can appear somewhat confusing at first, so it's well worth spending a few minutes familiarizing yourself with the interaction between these two layers of quantizing.

Figure 6.33
Quantize settings made in the Region Parameters box will automatically be visualized in the Piano Roll editor.

Start by recording a simple sequence into Logic and open this up in the Piano Roll editor. At this point, you should see the various imprecise placements of the notes to match what you originally played. Now move up to the Region Parameters, as part of the Tracks area Inspector, and set in Quantize setting to an appropriate value, like 1/16th. Glancing back at the edited part, you should now see the notes snapped exactly to the grid.

Now let's try quantizing from directly inside the editor itself. The advantage here is that we have choice over what notes are quantized – we could for example, only quantize a few notes that are played in a particularly sloppy fashion, or make a distinction between a collection of notes quantized on a 1/16 setting and another set of notes that should be played on a 1/16 Triplet (1/24) setting. Of course, settings applied in the Track area's inspector are applied wholesale to the entirety of the region, making it an unselective form of quantize.

We can apply quantize in from within the Piano Roll using two distinct methods, although both techniques make use of the Local Inspector which, if closed, can be activate vial the local menu item View > Show Local Inspector (of course, you can also use this menu item to hide the local inspector, if you wanted some more space on a cramped screen). The local inspector includes a number of options – including Time Quantize, Scale Quantize, and Velocity – that can be applied to notes selected within the Piano Roll editor. To quantize one or more notes, therefore, select a desired quantize setting from the local inspector (like 1/16, for example), highlight the desired notes and press the Q button next to the Quantize setting. Much like regions in the Tracks area, the notes remember their assigned quantize setting, so you can always select a note at a later point and reassign it accordingly.

Figure 6.34
Quantizing can also be directly applied within the Piano Roll editor, negating any exiting quantize settings that have already been set in the Region Parameters box. This allows you to quantize on a note-by-note basis.

As an alternative to having select notes and then assign a quantize setting, you can also use the Quantize tool. In this example, set the Quantize setting first, Select the Quantize tool, and then click on the notes you want to quantize. As you'd expect, the Quantize tool is a great way of flying through a MIDI performance and quantizing just a few notes as and when you see them.

Note, however, that with both techniques the settings with the Piano Roll editor ignore any settings made in the Region Parameters. However, should any settings be reapplied in as part of the Region Parameters and then the current quantize settings (in the Piano Roll) are lost.

With either level of quantize – Region-level or Piano Roll-level – it's important to remember that newer setting or modifications will overwrite existing quantize positions. So, for example, settings established in the Piano Roll editor will be overwritten if quantize is subsequently applied at Region-level, and vice versa. Operationally speaking, this is a logical solution, but it does mean that it's worth being clear where and how you're applying quantize.

6.16 Working with Controller Data Using MIDI Draw

Although the Piano Roll editor does an excellent job with note data, it doesn't quite provide the full picture with respect to MIDI controller data – including dynamic movements of modulation, pitch bend, volume, and pan data. However, by opening up the MIDI Draw view at the bottom of the Piano Roll window, you'll be able to see controller data as a series of lines with nodes along their length to edit the respective parameter levels. To open the MIDI Draw display, view select View > MIDI Draw (followed by the controller of your choice from the list) or open the window using the MIDI Draw icon toward the top of the Piano Roll window.

To better understand what can be achieved in the MIDI Draw area of Piano Roll editor, let's take a look at some typical editing tasks.

Editing Velocity in MIDI Draw

Although we can edit velocity data in the conventional note display of the Piano Roll editor, there are several operations – like a gradual rise or decay of velocity, for example – that can be better

Figure 6.35
Editing velocity using the MIDI draw is a much better way of drawing in a gradual rise or fall of velocity.

achieved using MIDI Draw. To edit velocity using MIDI Draw, you'll need to select the Velocity option from the drop-down list accessible from the small arrow icon to the left of the MIDI Draw area. Once the Velocity view is active, you should be able to adjust the individual velocities, although its most powerful feature is creating a gradual fall or rise over time. To do this, click and hold with the mouse, only releasing when you find a suitable "start point" for the line. Now, move the pointer over to end point, noticing the trail that is left behind it. When you click again, the virtual line will be drawn, effectively raising or lowering the velocities accordingly.

Editing Volume

Volume moves – like a number of other generic controllers including pan, modulation, and pitch bend – can also be drawn directly into the MIDI Draw area, allowing you to creatively shape an instrument's sound proprieties over time. In the specific case of volume and pan (assuming you're using it to control a software instrument), you'll actually see the volume and pan controls on the mixer move to reflect the curves you've drawn in the hyper view.

Drawing curves for these controllers can be done in two ways. The first technique is to use the pencil and effectively draw in the moves "freehand," although, in fact, Logic will actually place a series of node points along the duration of the part. The second technique is the pointer tool that allows you to place, edit, or remove the nodes one by one, which might be a better solution for more controlled movements.

Figure 6.36
MIDI Draw can provide an effective solution for drawing in more generic controller moves – like volume or pan, for example.

Editing Other Controllers, Including Filter Cutoff

If you can't see the controller that you need in the pop-up list, you can always use the Other option to access any 1 of the 127 different MIDI controller message types. For example, it might be that your software instrument or hardware synthesizer responds to MIDI controller message as a means of controlling filter cutoff, for example, or that it can be configured to work with certain controller message.

Figure 6.37
Use the Other option to display a
MIDI controller type outside the
main functions like volume and pan.

6.17 MIDI Draw in the Tracks Area

Although we're going to hop out of the Piano Roll editor for now, it's important to note the parity between MIDI Draw editing in the Piano Roll and the MID Draw facilities in the Tracks area. In effect, the Tracks area provides the full range of MIDI Draw features that we find in the Piano Roll editor all conveniently accessible via the View menu. Immediately, you should notice small previews of the moves we've made, but by selecting a region, choosing View > MIDI Draw, and selecting the controller type from the drop-down menu, we can see and edit the controller moves as if we were in the Piano Roll editor itself.

Figure 6.38
Using the View >
MIDI Draw menu
option, you can see
and edit moves that
you've created in the
Piano Roll editor.

As you'll see, the MIDI Draw feature is very close to the functionality of Logic's Automation system – something that we'll take a closer look at in Chapter 8. In particular, what we're playing here is a form of region-based automation – in other words, data and automation moves that are directly stored within the MIDI data of a region.

6.18   Going Further: The Piano Roll's Edit and Functions Menus

One of the quickest ways of speeding up the editing process is to explore alternative methods of selecting and processing note events. For example, you might only be interested in notes after a given position in the region, notes that have been previously muted, or notes at the top or bottom of a group of chords. By applying an automated selection process, you can achieve complex editing operations with relative speed and efficiency. Equally, various features within the Functions and Edit menu allow you to select and manipulate MIDI information in ways you wouldn't immediately think of, but is, nevertheless, a gold mine of useful creative options and technical solutions to MIDI editing.

Figure 6.39
The Edit and Functions menu facilitates powerful tools both for the selection and processing of MIDI data, making MIDI editing both more efficient and more creative.

6.19   Intelligent Selection: The Edit Menu

Besides the usual cut/copy and paste options, the Edit menu contains a number of different selection criteria. To better understand the Edit menu functions in the production process, though, let's look at some key musical applications of these features.

Select Equal Subpositions

Select Equal Subpositions is a great tool for working with any rhythmic regions where the material spans several bars. On a hi-hat pattern, for example, we might want to accent particular beats across a number of consecutive bars. Start by selecting the appropriate notes in the first bar of the sequence and then go to the Edit > Select > Same Subposition. Having activated the function, you should now see the appropriate notes selected, which you can then increase the velocity of using the velocity tool.

Figure 6.40
Here's our simple hi-hat pattern; in the first bar, we've selected the accent points that we'd like to pick up across all the bars in the region.

Figure 6.41
Using the Select Equal Subposition, we can add the corresponding notes in the subsequent bars quickly and easily. Now, all we need to do is raise the velocity to complete the effect.

Inverting a Selection

Invert Selection is a useful way of "toggling" selected notes in the Piano Roll editor. Say, for example, you're interested in modifying a large number of notes (maybe the unaccented notes in the example of the hi-hat), but at the same time, leaving a few notes unmodified. In this example, you'll achieve the results much quicker by selecting the notes that you don't want to modify, then inverting the selection (Shift + I) to toggle the selection, leaving the actual notes you want to modify selected.

Invert Selection is a great editing function to use in conjunction with the note-muting feature. For example, say you want to distribute note data between two virtual instruments from the same "parent" region. Starting with the first region, use the mute tool to remove the notes that you aren't interested in hearing with this particular instrument. Now copy the region over to another instrument track, open it up, and select the muted notes using the Edit > Select > Muted Notes. Unmute the selected notes (Ctrl + M), invert the selection (Shift + I), and then mute the newly selected notes. From one simple selection, you should have now made two regions with an alternate set of active notes.

Figure 6.42
Inverting and muting selections between different instruments can be a useful way of distributing MIDI information across your arrangement.

Select Highest Notes, Select Lowest Notes

By using the keyboard shortcut Shift + Up arrow or Shift + Down arrow, you can select the top and bottom line of a polyphonic region (in other words, a region with multiple notes at any given point in time). One immediate application of this is dividing a played chord of the keyboard between three or more monophonic string instruments (violin, viola, and cello, for example). To do this, simply copy the three-part harmony over to the various instruments and use the Select Highest Notes, Select Lowest Notes accordingly, along with the mute function, to remove the required notes in each part.

Figure 6.43
Use the Select Highest Notes or Select Lowest Notes to strip out the top or bottom line of a polyphonic MIDI performance. This could then be assigned to different instruments, for example.

6.20 Quick-and-Easy Note Modifications via the Functions and Edit menu

Both the Functions and Edit menu includes a number of different tools for both selecting notes and applying various transformations to MIDI data. The tools include some marvelous options for dealing with music arrangement, especially with orchestral samples – either to create fluid legato passages, for example, or split played chords out into a series of parts.

Note Trimming and Force Legato

Various options under the Trim submenu are interesting tools to use when you're working with instrument samples – like violin or clarinet, for example. Use Trim > Note to Remove Overlaps with Selected (assume you've selected the notes first, that is) on a MIDI region being directed to a "monophonic" acoustic instrument, like an oboe or clarinet. In the real world, it is impossible for these instruments to play two notes at the same time, so any accidental note overruns (where one note has been held for too long) can affect the realism of the output. Removing overlaps, therefore, will remove any accidental overlapping of notes.

Alternatively, Trim > Note End to Selected Notes (Force Legato) is a great solution where you want the part to take on a fluid, expressive quality – like violins playing a lead line. With a Forced Legato, all the notes run into each other, irrespective of their original played length, producing a smoother overall line.

Figure 6.44
Using the Force Legato feature – Trim > Note End to Selected Notes (Force Legato) – this MIDI region has had all its notes lengthened to meet the subsequent note.

Voices to MIDI Channels

Use the local menu option Functions > Set MIDI Channel to Voice Number feature as an intelligent, automatic means of splitting a polyphonic chord part into a series of unique voices. This end region will look the same, although on closer inspection you'll find that each voice has been assigned a different MIDI channel. To strip the data out, go back out to the Tracks area, Ctrl click the region and choose MIDI > Separate by MIDI Channel. The polyphonic region should now be exploded into a number of different regions, one for each "voice" of the original part.

197

Deleting Duplicates

Now and again it's possible for two MIDI events, of the same pitch, to end up on top of each other. This usually happens as a by-product of quantization, and, if you listen carefully, can be heard as a slightly "airiness" or "phasing" on the note in question. If you suspect this is happening, one quick way of rectifying this is to use the menu option Edit > Delete MIDI Events > Duplicates option from the local menu. Although you might not see any immediate difference, Logic will have deleted these "double" events, with the note sounding a more natural way.

MIDI Transform

One of the most interesting and useful elements of the Functions menu is the MIDI Transform options. The MIDI Transform operations provide a number of radical, but nevertheless useful ways of modifying data that would be time-consuming to apply on a note-by-note basis. Options like Fixed Note Length, for example, speak for themselves, setting all note values to the same duration. The Fixed Note Length option is a great choice if you're trying to transform an existing "played" performance into a sequencer-like performance, with rigid note lengths for each step. Another useful option under the MIDI Transform section is Humanize, which takes an existing performance and randomizes the position, length, and velocity of each note, adding the subtle "imperfection" of a human performance.

Figure 6.45
The MIDI Transform menu includes a range of unusual tools for MIDI editing.

KNOWLEDGEBASE 2
Using Multioutput Instruments
Both the EXS24 and Ultrabeat instruments are available in multioutput versions, in addition to their standard mono and stereo configurations. It's also possible that a number of other third-party software instruments in your collection – like FXpansion's BFD – might also come in multioutput versions. The notion of a multioutput is simply to provide additional outputs from the plug-in.

(Continued)

Each output has its own fader, and, therefore, the possibility of adding additional effect plug-ins specific to the output in question, rather than the instrument as a whole. This is most often the case with drums, where you might wish to place different reverb, EQ, and compression settings on each different part of the kit.

To access the additional output, you'll need to ensure that you instantiate the specific multioutput version of the plug-in. To add in additional aux channels, simply click on the + sign at the bottom of the instrument channel strip, with each

Figure 6.46 Use the multioutput versions of instruments like the EXS24 and Ultrabeat so as to be able to process and mix individual elements from their sound output.

subsequent press adding another output into the equation. To access the outputs, you'll need to route the appropriate signals using the plug-in's controls. On the EXS24, for example, the routing can be achieved via the EXS24 editor, while Ultrabeat's output options can be found next to each drum voice.

6.21 Step-Time Sequencing

Step-time sequencing dates back to the earliest analog CV sequencers that wouldn't technically allow you to record real-time musical information (as Logic does), but instead store music as a series of 16–32 control voltages. Nowadays, however, step-time sequencing is still a valuable tool, mainly as a means of "stepping through" and creating a MIDI sequence piece-by-piece, rather than having to play a part by hand. If you're creating sequencer-driven music – with flowing arpeggiators and fast-moving basslines – then step-time sequencing arguably offers a more effective means of recording a part, without the need to quantize and meticulously edit your original sloppy performance.

Figure 6.47
Here's an example of sequence created in step time, with a fast-moving sixteenth arpeggio effect.

To engage step-time recording in the Piano Roll editor, you'll need to activate the small MIDI In icon in the top left-hand corner of the window. Whenever you press a MIDI note in step-time mode, Logic will place one or more (if you play a chord) notes at the current song position and then advance one step forward at an amount defined by the current Snap setting. Working in this way, you can quickly create a fast-moving sequencer line playing each note in sequence, but without having to play in time with the metronome.

As an alternative to using a MIDI keyboard, you can also make use of Logic's Musical Typing Keyboard (a feature that allows to play an instrument using your Qwerty keyboard, which can be called up via the menu option Window > Show Musical Typing) as well as the dedicated Step Input Keyboard found under Window > Show Step Input Keyboard. The Step Input Keyboard is a good solution for drawing note events using the mouse, with a clear horizontal representation of the musical keyboard, selectable note lengths, velocity settings, and so on.

Figure 6.48
Use the Step Input Keyboard and Musical Typing Keyboard as alternative methods of note entry for step-time recording.

6.22 The Step Editor

Although we've spent some detailed time exploring Logic's main MIDI editor – the Piano Roll – it's worth familiarizing yourself with some of the other MIDI editors and what they can individually offer to the MIDI editing process. The Step Editor has some comparison with MIDI Draw in that both systems are primarily designed as means of editing controller information – like volume, pan, or modulation. The specific benefit of the Step Editor, though, is that we get to see a number of track lanes at the same time.

Editing and Creating MIDI Controller Data in the Step Editor

One quality that should become apparent about the Step Editor is that it prefers to display its controller data as a series of 1/16th steps (hence the name!), rather than the line + node approach of MIDI Draw.

Technically, this creates a slightly different approach to editing, although the end effect is much the same. It does, however, also make for an excellent source of pseudo-random sixteenth Sample-and-Hold movement, where a pattern appears to pulse up and down in time with the music. Try doing this with the modulation controller, for example, mapped to filter cutoff on the ES2 synthesizer.

For more generalized movements, you can use the pencil tool to draw in a series of stepped MIDI events – maybe creating a quick ramping up and down of volume, for example. Using the line tool, you can also draw more gradual movements up and down, in the same way as the line feature in MIDI Draw.

Figure 6.49
The Hyper Editor
is particularly
good at stepped
sixteenth controller
movements.

Adding New Controller Types and Lane Sets

To add a new controller type other than the ones in the preexisting set, simply double-click in the blank space beneath the last controller assignment, effectively creating a new controller track lane. Opening up the Inspector, you can then define parameters in relation to the controller lane, including the pen width, grid, and most importantly, the MIDI status type (program, controller, and so on), controller number (if applicable), and MIDI channel.

The particular arrangement and order of controller track lanes can be saved as part of your own, custom-designed Lane sets. You could, for example, create a Lane Set that specifically relates to the MIDI control of a particular instrument plug-in or a simplified set based on mixing. To create or delete a Lane Set, go to the Lanes menu and select the corresponding option, or alternatively, use the dedicated Lane Set selector (where you can also switch sets) embedded into the Inspector.

KNOWLEDGEBASE 3
Separating MIDI Events

The Separate MIDI Events feature is a simple way of exploding a MIDI region into its constituent parts – this could be based on note pitch, for example, or by the Event Channel. To Demix a region, first select it, then choose MIDI > Separate by MIDI Channel or MIDI > Separate By Note Pitch from the contextual menu. The note pitch option, for example, is a good solution to use in combination with a MIDI drum loop, where a number of different lines (hi-hat, Toms, and so on) have been sequenced into the same region. By separating the MIDI region, you'll create individual parts for each part of the kit, which you can then label Kick Drum, or Snare, for example, accordingly.

The Event Channel option works well combined with the Functions > Set MIDI Channel to Voice Number feature in the Piano Roll, where a polyphonic part has been split into a number of different MIDI voices.

Figure 6.50
Demixing allows you to explode a single region into a number of separate parts based on note pitch or MIDI channels.

Editing Drums Using the Step Editor

In addition to editing controller data, the Step Editor can also be used to edit and create note data, using the same series of sixteenth "blocks." As such, the Step Editor is a good solution for drum programming, with the height of each block providing quick-and-easy velocity adjustment, while the pencil tool allowing to draw in straight lines of sixteenth events. Usefully, Logic provides a preprogrammed GM Drum Kit Lane Set, specifically designed for drum programming. As most drum kits are mapped to the GM standard, this should be good for the majority of applications, although you can always program your own drum-based Lane Sets for plug-ins like BFD or Battery.

Figure 6.51
Use the GM Drum Kit Lane Set as a useful way of creating and editing drum sequences. Note that you can also create your own Lane sets to match the mapping of virtual instruments like BFD.

6.23   Score Editor

The Score Editor is the final MIDI editor selectable from the row of tabs in the editor pane. Obviously, this editor is an appropriate choice if you have an ability to read printed music, although it has several key limitations for serious and detailed MIDI editing tasks. As a score is effectively "visually" quantized, the Score Editor provides us with a relative coarse view of a note's placement and, while we're at it, its length. This makes any change beyond pitch manipulation somewhat flawed.

Of course, the real benefit in including the Score Editor is in the preparation of a score for musicians to play. Music preparation, as it's called, is a process we'll be taking a closer look at in Chapter 11.

Figure 6.52
The Score Editor is a quick-and-easy way of editing pitch-based information, but provides us with little control or information on positional data.

6.24 Event List

The last MIDI editor that we're going to take a look at is the Event List editor, which displays the MIDI information in pure "data" form. The Event List can, at first, feel a little daunting and perplexing, and certainly not comparable to the visually driven approach to MIDI editing defined in the Tacks area, Piano Roll, Score, and Step Editor. However, there are times when it's important to get "down-and-dirty" with the data – understanding precisely where an event occurs, for example (down to tick-level accuracy), or the exact parameter level with a controller.

The Event List editor can be opened as part of Logic general "list" displays – including markers, tempo, and time signature lists – that can be opened up on the right-hand side of Logic Pro's main window. This is an effective solution, as often, you'll want to combine the Event List with other MIDI editors like the Piano Roll or Hyper Editor. To open the Lists display, press D on your keyboard, or View > Show List Editors, with the Event List accessible from the Event tab at the top of the display.

Figure 6.53
The Event List provides an alternative text-based view of our MIDI data. Combined with the other editors, the Event List can be an indispensable tool to accurately understand the content of your MIDI region.

Filtering Events

Given the range and quantity of MIDI data that can be included in a region, it is often necessary to filter different types of MIDI data out of the Event List. By clicking on the Filter tab, you can then switch in and out various MIDI events, including notes, program change, controller, and so on. Using the filter can also be an interesting way of selectively deleting certain types of MIDI data. For example, you could filter note data so as to select and then delete extraneous MIDI data (controllers and so on) that you might not be interested in.

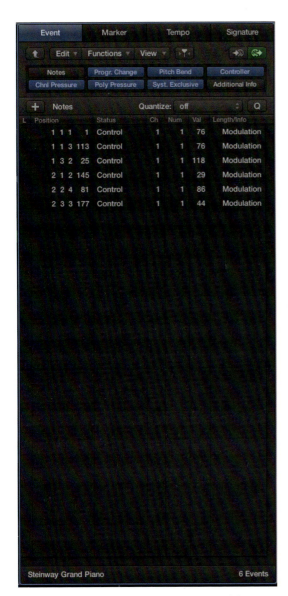

Figure 6.54 Use the Filter tab as means of filtering out certain types of MIDI data.

Modifying and Adding Events

Editing data in the Event List is as simple as clicking on the appropriate number and either dragging up and down or double-clicking and entering a new value numerically. Numerical entry can be a surprisingly effective way of positioning notes and other MIDI events both quickly and accurately. You can also add new events, based on the current song position and preassigned message type, by clicking on the + icon. For program changes, or static controller levels, the Event List shines as a quick effective solution with the minimum of fuss.

6.25 Creating Rhythm Tracks with Drummer

Unlike sequencing with other virtual instruments like the EXS24, the combination of Drummer (and its associated plug-in, Drum Kit Designer) needs special attention. In effect, "Drummer" is a completely different type of track (different from Instrument Tracks or Audio Tracks) that needs a unique approach. The good news, however, is that Drummer is specifically designed to make rhythm programming a speedy and intuitive task, which is certainly not the case if you are programming from scratch using a virtual instrument.

Creating a Drummer Track

As we've seen in the "Starting a Project" chapter, a Drummer track can be created via the New Track command. When you create a new Drummer track you'll notice a couple of important things. First, rather than starting from an empty track, Logic will have created two default Drummer regions, distinguishable by the unique series of transient-like shapes that indicate the position of drum hits. In addition to this, Drummer also gets its own editor window found in the same position as the Piano Roll editor, which allows you to edit parameters on a region-by-region basis. Finally, Drummer also instantiates its own instrument – Drum Kit Designer – which we've introduced in Plug-in Focus 6.

Figure 6.55
Creating a new
Drummer track will
load an instance
of Drum Kit
Designer as well as
opening the special
Drummer editing
window.

Basic Rhythm Creation with Drummer

What's unique about Drummer is the fact that it uses "algorithmic" techniques to create a rhythm track, rather than requiring you having to program each and every part of the drum pattern. In short, it works like a drummer – you instruct a musician to play in a certain style, with a degree of complexity and overall playing dynamic (loud or soft, in other words), and they come back with a pattern or groove for you to work with. If it isn't right, you simply ask them to play in a different way, whether it's a little quieter, for example, or with more ghost notes.

The concept with Drummer, therefore, is that you can select a number of key criteria – the choice of drum kit elements, simple/complex, loud/soft, and so on – and Logic will then create a rhythm pattern in accordance with this. This might sound similar to preset bashing, or simply picking a loop from a collection of different patterns, but it actually belies the complexity of what's happening "under the bonnet." In truth, what Logic outputs is an intricate human-like performance, complete with the subtle sonic variations, fills and ghost notes that a real drummer would produce.

Figure 6.56
Drum patterns are
programmed using
simple musical criteria

PLUG-IN FOCUS 6
Drum Kit Designer

Drum Kit Designer is Logic's virtual drumming solution, and is intended to be paired with a Drummer track (which we'll explore in Section 6.25) although of course, you can use it with any form of MIDI Sequence.

Opening up the plug-in reveals a pictorial version of the kit, which you can then edit and reassemble as you see fit. The preset options offer various default kits to get you started (Birch Kit, Classic Sixties Kit, and so on), although you can also click on each kit element individually and exchange parts of the kit to build up your own hybrid kit – maybe pairing a Seventies Felt kick, for example, with a tight Modern Chrome snare. Each Kit Element also has its own series of parameters, including Tuning, Dampening, and Gain, which further allows you to tailor the sound as you see fit.

Alongside the main stereo kits, it's also worth noting the Producer Kits (found via the Library), which present the drums as a Track Stack. Although the Producer Kits are more memory intensive, they do allow you to access each and every part of the kit as an individual channel strip, facilitating much greater freedom in how the kit sounds.

Figure 6.57 Drum Kit Designer lets you assemble and fine-tune your own custom virtual drum kit.

To edit a Drummer performance, click on one of the two Drummer regions and then turn your attention to the editor at the bottom of the screen. Starting from the top left-hand corner of the interface we can specify the broad genre group we're working with – whether it's Rock, Alternative, Songwriter, or R&B. From within the Genre you'll then be able to select a Drummer, each with their own musical personality traits and a choice of drum kit.

Once you've loaded your choice of Drummer, you can then start to play with the pattern itself. In essence, what they're playing is defined by the X-Y pad, the choice of drum kit elements you want to work with, and the position of a number of parameter slides and dials. The overarching performance is defined by the X-Y pad – whether you want the pattern soft and simple, for example, or complex and loud. Next comes the choice of kit elements – whether drummer is using the kick, snare, and cymbals, for example, or just the snare and toms. Alongside each kit element is a slider, used to define the musical pattern player by that part of the drum kit – like a basic offbeat snare, for example, or a 1/2 time groove.

The two final controls govern the relative amount and density of fills as well as the amount of Swing applied to the pattern. The relative position of all of the pattern controls can be then stored and recalled as part of the presets system found toward the left of the X-Y pad. The presets system is also a good way of quickly auditioning some radically different sounding grooves, so it may well be a good place to actually start your rhythm programming.

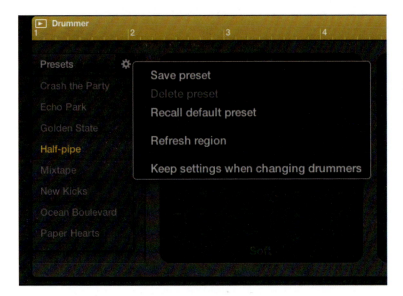

Figure 6.58
Use the Presets feature to save or load pattern information that you've specified using Drummer's controls.

How Regions Work with Drummer

If you haven't clocked it yet, the region is an important part of the Drummer concept. First, each region has its own set of pattern controls (not the choice Drummer, of course, but the X-Y pad positioning, choice of kits elements, and so on). You'll also notice elements like the fills are contoured so that they tally with the length of the region, with the fill only occurring over the last bar or so of the region.

Figure 6.59 With Drummer, each region has its own set of controls and therefore, a unique pattern.

Given this behavior, it makes sense that you make a series of regions that confirm to the overall structure of your song (verse, chorus, middle eight), or where you want the drums to change. Use the pencil tool to create new regions and resize them using the pointer tool.

Advanced Drummer Techniques

Delving deeper into Drummer, it's worth noting some of the more advanced controls as part of the Details section and on the front page of the editor interface. Checking the Follow button, for example, instructs Drummer to adjust the pattern in relation to other musical events – locking the drums to a bass guitar, for example, or making the kit tightly follow a drum machine pattern.

Over the page, as part of the Details section you'll find control for Feel, Ghost Notes, and the Hi-Hat. Feel nudges the pattern forward or backward in time – Push adding an urgency to the pattern by the notes being played ahead of time, while Pull lets the groove sit back a little, giving it a more relaxed quality. Ghost Notes add a form of "color" to the loop, with the drummer playing softer notes in between the main beats. The Ghost Notes control sets the relative level of the drum hits, so the effect is either discrete (with Quiet notes) or a more noticeable shuffle, where the Ghost Notes are played louder. Finally, Hi-Hat can force the pattern to move between a closed and open articulation, or simply leave it on automatic to have it dictated by the pattern's existing setting.

Figure 6.60
Use the Follow option to have Drummer intelligently match the pattern to a part in your arrangement.

Figure 6.61
Use the Details page to refine your pattern further.

Loading a Producer Kit

Adjusting Drummer's performance is one part of the eventual result, but we also need to consider the importance of the sound produced by the kit. As we explored in Plug-in Focus, Drum Kit Designer is the sonic heart of Drummer, and the main place to explore if you want to swap-out parts of the kits, (like swapping the snare drum, for example), or changing aspects like the Drum's tuning. In addition to this though, we might also want to consider how we can use signal processing plug-ins – like compression and equalization – to change to the sound of the kits.

By default, Drummer works with "stereo" Drum Kit Designer patches – in other words, the kit is largely premixed and simply routed to the stereo output. The so-called Producer Kits, though, facilitate access to each and every part of the kit, so that you can hear and modify the individual kick mic, for example, or listen to just room mics in isolation. In short, the Producer Kits give you the ultimate in sonic flexibility over the kits, with the same degree of freedom as if you'd recorded the drums yourself.

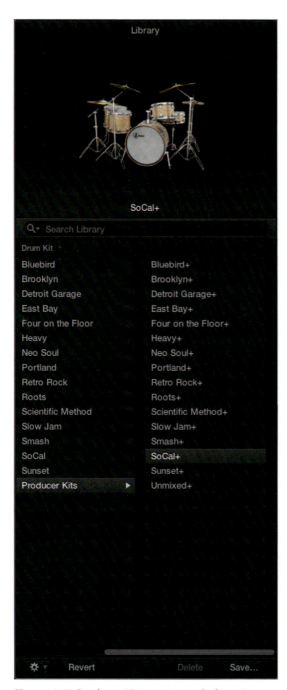

Figure 6.62 Producer Kits are expanded versions of the main stereo kits, with individual access to each and every part of the kit.

You can find the Producer Kits via the Library browser, with a discrete folder called Producer Kits making them easier to find. The name of the Producer Kits matches the main stereo kits, although the addition of a "+" suffix makes it easy to discern the difference. Once you've loaded the Producer kit, click on the small arrow on the accompanying track lane to expand the Track Stack. With the Track Stack expanded, you see each and every part of the kit, allowing either to simply rebalance the mix as you see fit, or go even deeper and process each part of the kit in a different way (which is something we'll cover in more depth in Chapter 8).

Using Drummer with Other Virtual Instruments

It would be easy to assume Drummer is a closed off system, but it is possible to use the Drummer performance with an alternative virtual instrument. Start by creating the basic groove using the Drummer track and editor; once you're finished, though, create a new virtual instrument track and copy the Drummer region over to the new track. As you move the region you notice it change slightly, from the transient-like shapes used by drummer to the more familiar "blobs" of MIDI data. You won't be able to edit the performance again using the Drummer interface, but you will have its output as raw MIDI data.

Given that Drummer exports the performance using the GM drum mapping standard, you should find that you can use the MIDI triggers with a range of other virtual drumming plug-ins, like BFD or Native Instruments' Abbey Road Drums series (assuming you set them to an appropriate GM drums mode). Interestingly, you can also use many of Logic's Drum Machine patches (found via the Library) in the same way, which can also be useful if you want to double Drummer's output with something more hard hitting.

Figure 6.63 Expanding the Track Stack facilitates access to the original mic channels used to record the kit.

Figure 6.64 Copy the Drummer region across to a new track will convert the sequence data into MIDI.

6.26 Working with MIDI Plug-Ins

Having explored a number of traditional ways of editing MIDI Data and controlling virtual instruments, let's look at a completely radical approach – MIDI plug-ins. Rather than working with the raw data of a performance, MIDI plug-ins work as "live modifiers" instantiated at a point between your original performance data and the virtual instrument used to playback the musical information.

Arguably the best example of the "live modification" is an arpeggiator, which can be traced back to the days of analog synthesizers and CV sequencers. An arpeggiator works by analyzing your playing (which, in effect, forms the "input") and then outputting a series of stepped messages at precise tempo-locked positions. Play a simple three-note chord, therefore, and an arpeggiator could produce swathes of 16th notes spanning three or more octaves, or alternatively, it could produce a chugging rhythmic sequence using the three notes played in a repeated fashion.

MIDI plug-ins are inserted directly into the audio mixer for the corresponding software instrument. Simply click in one of the empty slots beside the MIDI FX legend and pick from the selection of MIDI plug-ins. Just as with signal processing plug-ins for mixing, you can insert and instantiate a number of different MIDI plug-ins each adding to the results from the last plug-in in the signal path. Plug-ins can be deactivated by clicking on the left-hand side of the insert slot, or picked up and reordered.

Figure 6.65
MIDI plug-ins are placed directly into the signal path of the audio mixer, under the MIDI FX slot.

The MIDI plug-in will process all information routed to the software instrument, while the original performance data remains untouched. Just as with audio processing, bypassing the plug-in will restore the MIDI performance data to its original form.

6.27 Types of MIDI Plug-Ins Included in Logic Pro X

Arpeggiator Plug-In

The arpeggiator transforms one or more MIDI notes into a tempo-synced arpeggiating pattern. Parameters include the ability to define the motion of the pattern (Up, Down, Up & Down, and so on), and the amount of octaves that it works over. A unique pattern mode can even turn the Arpeggiator into a form of CV step sequencer, which is particularly great mapped to parameters like filter cutoff. We'll look at the Arpeggiator in more detail in Section 6.28.

Figure 6.66
The Arpeggiator
Plug-in.

Chord Trigger Plug-In

The Chord Trigger plug-in lets you trigger a user-defined chord from just a single note. In Single Mode, a single user-assigned chord tracks the keyboard fully, which is an interesting way of creating parallel chord movement, especially if your keyboard technique isn't that good! Multi Mode is interesting in that it permits a unique user-defined chord for each note, allowing you to create some complex chordal movement from just a few simple note triggers.

Figure 6.67
The Chord Trigger
Plug-In.

Modifier Plug-In

Use the Modifier plug-in as a means of reassigning MIDI controller information on-the-fly. For example, if a virtual instrument only permits control of its filter on a specific controller, simply remap this to the modulation wheel (using the Input Event and Reassign to parameters). The Scale and Add options are useful if you want to adapt previously recorded controller movements, using the Scale parameter, for example, to make the movement more or less dramatic, or the Add control to simply add or subtract to the existing values.

Figure 6.68 The Modifier Plug-in.

Modulator Plug-In

The Modulator is an interesting plug-in because it can actually create new MIDI controller data in response to note information fed to its input. To understand the Modulator Plug-in it's best to pair it with an instance of the Retro Synth, where the plug-in adds an additional layer of modulation possibilities. That said, the Modulator Plug-in can also work wonder with a range of virtual instrument, especially those that map the modulation wheel in an expressive way.

Figure 6.69
The Modulator Plug-in.

Figure 6.70 The Note Repeater Plug-in.

The Modulator plug-in contains two distinct sections – an LFO and an Envelope. The LFO, for example, can be set to "Free" mode, outputting stepped 16th controller movements in time with the track's tempo. Alternatively, set it to Single mode, and the LFO works like a part of the synthesizer, triggered as you play an incoming note. The Env section works in a similar way to the LFO in Single Mode, triggered from note played on the keyboard, although you can also trigger in response to the LFO, which is great for tempo-sync'ed envelope movements.

Note Repeater Plug-In

The Note Repeater plug-in is the equivalent of a MIDI delay line, outputting delayed note messages in respect to its input. Specify the number of delayed notes using the Repeats parameter, with the delay time established with the Delay setting (which can be set to sync with your tempo, of course). Creative treatments can come from experimentation with the Transpose and Velocity Ramp. Use a +1 Transpose and Velocity Ramp to 73% to create an upward-spiralling effect where each repeat is a little quieter. It's also well worth pairing the Note Repeater with the Chord Memorizer (placing the Note Repeater after the Chord Memorizer) to hear how two MIDI plug-ins can work together dynamically.

Randomizer Plug-In

As the name suggests, randomizer can be used to randomize note, velocity, pitch bend, or controller data. Arguably its best application is in creating small randomization to velocity for drum sequences that need to mimic a human performance.

Scripter Plug-In

Not for the fainthearted, the Scripter plug-in lets you code your own MIDI plug-ins using JavaScript. Some example presets give you a taste of what you can achieve, and it will be interesting to see if any third-party developers make use of the scripter plug-in in relation to the use and application of sample triggering.

Figure 6.71 The Randomizer Plug-in.

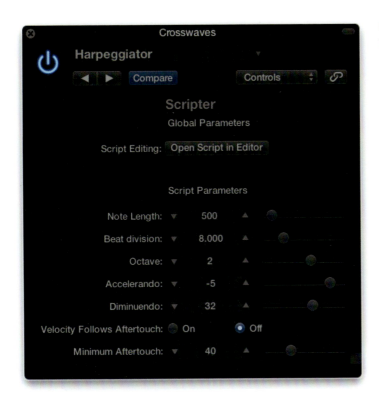

Figure 6.72
The Scripter Plug-in.

Transposer Plug-In

As well as the oblivious transposition activities, the Transposer plug-in is also interesting as a means of applying scale correction, so that outputted MIDI notes correspond with a predefined musical scale. The obvious application here could be in relation to rectifying poor musical technique, but it's also interesting to think about possibilities of combining the Transposer with the Randomizer plug-in, especially in examples where notes values are randomized but then brought back in-line with the Transposer Plug-in.

Velocity Processor Plug-In

Think of the Velocity as a form of MIDI compressor/expander, either used to restrict the range of MIDI velocities within your performance, or expand them. The Velocity Processor Plug-in works particularly well with sampler instrument that used velocity-switched samples, like Drum Kit Designer or some of the more expressive instruments in the EXS24 library. The plug-in works like a compressor, with a threshold and ratio control (for more information on compression see Plug-in Boxout 3 in Chapter 8) as well as gain makeup.

Figure 6.73 The Transposer Plug-in.

Figure 6.74 The Velocity Processor Plug-in.

6.28 Using the Arpeggiator

The Arpeggiator is arguably the most musically useful MIDI plug-in, so it's worth detailing a few practical examples with it.

Basic Arpeggiator Effects

The core musicality of the Arpeggiator is defined in the central Note Order section. First, define the Rate to set the speed of the Arpeggiator effect on musical divisions – like 1/16th notes, for example. The subsequent controls govern the precise shape of the movement, which includes options for playing the arpeggio with an upward or downward movement, as well as options to combine both upward and downward movements in the same pattern. Finally, use Oct Range so set how wide the arpeggio is in respect to octaves.

Figure 6.75
The Note Order controls establish the key musical information in relation to the Arpeggiator effect.

6.29   Using the Grid

The Grid adds another dimension to the Arpeggiator effect by letting you specify a unique velocity for each step of the Arpeggiation. Played with just a single note, the effect is similar to that of a CV sequencer, particularly where note velocity is routed to filter cutoff.

To activate Grid mode click on the Grid legend. You can then specify a unique velocity setting for up to 124 steps. The chord button beneath each step allows you to move the Arpeggiator between playing the next step of the Arpeggio (as specified by Note Order) and playing the notes as a chord. Used carefully, the Grid can create some wonderfully complex sequencer-like effects, using either a single note as an input or full chords.

Figure 6.76
Use the Arpeggiator's Grid to create CV Sequencer-like effects.

6.30   Advanced techniques

Many of the advanced controls for the Arpeggiator can be found under the Options tab. The Note Length is well worth exploring, allowing you to shorten or lengthen each note of the sequence. Short, staccato sequences, for example, can be really effects when you reduce the Note Length beneath 50%. You can also inject a degree of Randomness both to the aforementioned Note Length and also Velocity, which is a great way of creating random Sample-and-Hold movements. Finally, use Swing to create a swung arpeggio, which can be a good way to loosen an otherwise robotic sequence.

Figure 6.77
The Options tab contains many of the Arpeggiator's advanced features.

KNOWLEDGEBASE 4
EXS24 Data Management

Although it's quite possible to operate and work with Logic without an understanding of the EXS24's occasionally complex means of data management, it can become increasingly important as your own user-installed collection of sampler instruments begins to swell. The key to understanding where your data is, and how this is managed, is in differentiating the different sources of sample instruments – in other words, original library data installed with Logic, user-installed and created content, and data that specifically relates to the current project. Although this can seem complicated at first, a little time investment will allow you to better access, organize, and archive this important data.

Factory content that was installed with Logic can be found under the path Library/Application Support/Logic/Sampler Instruments, with the associated sample files (remember an instrument file is separate to the sample data) under Library/Application Support/Logic/EXS Factory Samples. Your data will be saved in your user file under *User Name*/Library/Application Support/Logic/Sampler Instruments. Whenever Logic is loaded, it will scan these two folders and build up an Instrument list based on the contents. This also means that you can actively use the Finder to organize the instrument files as you see fit, creating custom list of folders, for example, or even moving or duplicating favorite instruments into a single folder.

In addition to these two main folders, Logic will also take a look inside your Song folder to see if any EXS instruments are stored there. This may well be the case if you've inherited a project from another user who's backed-up a complete set of assets with their project archive. It should be pointed out, though, that these project-specific instruments cannot be accessed in other projects. To do this, you'll need to copy the instrument files into the appropriate Library folder.

Samples for any of these instruments can reside on any drive, in any folder, with Logic scanning for the appropriate samples as it loads the instruments. The two immediate implications here, though, are both the scattershot sample organization that can occur and the speed implication of a fragmented collection of samples. If possible, therefore, try to organize a large part of your sample data (you can leave the EXS Factory data) into a neatly separated folder, partition, or even better, a separate drive.

(Continued)

Figure 6.78 You can use the Finder to actively manage the organization and arrangement of your sampler instruments and how these appear in the EXS24 instrument menu.

In this chapter

Creative Sound Design

7.1 Introduction

For anybody actively engaged in the more creative aspects of designing and modifying sound – say, a film sound designer, for example, or a musician working in the realms of electronica – there can be little doubt that Logic represents one of the most exciting applications to work in. Right out of the box, the modern-day sound designer has access to a huge range or synthesizers, samplers, and signal processors without having to go anywhere near any third-party products. Indeed, it's highly likely that your whole sound design project (whatever form it takes) could be realistically achieved entirely within the realms of Logic.

Having worked on the basics of using virtual instruments and MIDI in the last chapter, this chapter goes on to look at the creative applications of Logic's virtual technology to shape and craft any number of interesting sounds – whether you're using its versatile set of synthesizers (ES2, Sculpture, and so on), the pure sonic muscle of the EXS24, or the range of esoteric signal processors. Most important, however, making your own sounds (rather than just simply relying on presets) can be a real opportunity to stamp your own identity onto a particular audio product, and ultimately, make your work stand out from the crowd!

7.2 Logic's Synthesizers

Beyond some of the basic synthesizers we looked at in Chapter 6 – the ES E, ES P, and the ES M synthesizers – Logic also includes a number of other synthesizer plug-ins, each dedicated to a different branch of synthesis and capable of an enormously wide palate of potential sounds. So, let's take a look at the three remaining synthesizer plug-ins, and what particular strengths they offer to the sound designer.

ES2

The ES2 is Logic's definitive answer to the good-old polyphonic analog subtractive synth – like Roland's Jupiter 8, for example, or Sequential Circuits' Prophet 5. However, this only forms part of the picture, as the ES2 has plenty of further tricks up its sleeve, namely, vector synthesis, wavetable synthesis, and a host of other sound design tricks borrowed from many of the classic digital synths of our time: the Waldorf's Microwave, the Korg Wavestation, and the Prophet VS. Make no mistake, this a tremendously versatile synth – not the easiest of plug-ins to get your head around, but one that will consistently surprise you with what it has to offer.

Figure 7.1
The ES2 – arguably Logic's most powerful and sonic dexterous synthesizer.

Retro Synth

The Retro Synth shares many sonic traits of the ES2, but its simplified interface and more refined feature set arguably make it a more approachable instrument to use. Arguably, Retro Synth's most distinctive is the series of four different modeling modes – Analog, Sync, Table, and FM – that mimic various vintage

Figure 7.2
Retro Synth's simple controls and powerful sound make one of Logic's finest synthesizers.

synthesizers from the 70s and 80s. Despite its simple controls, Retro Synth can produce some powerful and sonically effective sounds that are well worth closer inspection.

EFM1

The EFM1, although technically part of Logic's more budget-orientated synths, is actually an interesting and valid example of FM synthesis in action. For those of you who remember, FM synthesis was the primary system for synthesis used by Yamaha in the 1980s and was at the heart of their infamous DX7. Although technology has moved on significantly since the golden days of FM, it can still be a valid source of sounds impossible to achieve elsewhere.

Figure 7.3
The EFM1 is a simple synthesizer designed to exploit the digital wonders of FM synthesis.

Sculpture

Sculpture is easily the most cutting-edge synthesizer in Logic's collection and represents one of the most exciting developments in synthesis in the recent years – that of Component Modeling. Unlike synths like the ES2 and EFM1, Sculpture works on the basis of a series of complex mathematical models describing the precise acoustic behavior of a particular instrument. For example, rather than synthesizing the sound of a guitar by selecting a "similar" waveshape and approximating its behavior with filters and envelopes, Sculpture precisely predicts what happens as a string is plucked (using a variety of different plucking mechanisms) and what happens as the sound dies away. Not surprisingly, therefore, Sculpture produces sounds impossible to achieve on other synthesizers.

Figure 7.4
Sculpture is a unique software synthesizer dedicated to Component Modeling synthesis.

7.3   Understanding the ES2

For most conventional music and sound design activities, the ES2 will probably be your main synthesizer of choice, largely because of the range of subtractive synthesizer sounds it offers that are so eminently useable in electronic music. Although the ES2 isn't the easiest of synthesizer to

Figure 7.5
The ES2 can separated into five distinct sections – Oscillators (A), Filters (B), Amplifier and effects (C), Modulation Matrix (D) and LFOs and Envelopes (E).

understand, it's a great foundation for other instruments like Retro Synth and subtractive synthesis in general.

The full interface itself is divided into the five key areas, illustrating the principal audio signal flow through the synthesizer (Oscillator > Filter > Amplifier, from left to right) alongside the interaction of modulation sources like low-frequency oscillators (LFOs) and ADSR (Attack, Decay, Sustain, and Release) envelope generators. An additional "Macro only" mode – selectable from the bottom of the ES2 – presents a minimized control panel, which might be suitable if you're only interested in tweaking presets.

The Oscillators (A)

Toward the top left-hand corner to the interface, you'll find the controls for the three oscillators included within the ES2. The oscillators form the starting point of sounds created on the ES2 – the raw, harmonically rich waveshapes from which the filter gauges out harmonics, and a big part of the identity of the sound. Thankfully, the ES2 includes a number of different principal waveshapes, and even a wavetable (more on this later) as well as a unique triangular vector mixer for the purpose of blending the oscillators together.

KNOWLEDGEBASE 1
Waveshapes
As you've seen with the oscillator section of the ES2, waveshapes can have a big effect on the type of sound you can produce. However, what defines the sound of each waveshape, and how best can they be used?

Of the four principal waveshapes, the sawtooth produces the richest amount of harmonics, and is, therefore, the most popular choice for creating many of the classic synthesizers sounds. From warm pads to gutsy synth basslines, the sawtooth wave can really excel, giving the sound plenty of presence, depth, and bite! Be warned though, overuse of the sawtooth can make a track appear rather stoggy and one-dimensional, so it's worth exploring other variants and the sounds they can produce.

Compared with the sawtooth, a square wave produces a distinctive hollow and open sound to it, by the virtue of the fact that it only contains odd harmonics from the harmonic series (a sawtooth contains both odd and even). Acoustically speaking, this makes the square wave similar to the tone of an instrument like the clarinet; although it's also useful for bass sounds where it can sound "big" without grabbing large parts of the sound spectrum (and eating up your mix, therefore) like the sawtooth.

(Continued)

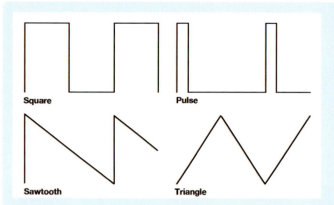

Figure 7.6
The Square, Pulse, Sawtooth and Triangle waveforms are the building blocks of subtractive synthesis.

The pulse wave, a variation of the square wave, is created by changing the duty-cycle of a square wave, effectively producing a wave that both sounds and looks thinner and more nasal than the square wave. For example, the ES2's Oscillators 2 and 3 both feature a fully variable square wave, allowing you to morph from a conventional hollow square wave to the thinnest of pulse waves. One common technique is to modulate the wave's width using an LFO, producing an effect otherwise known as pulse-width modulation (PWM). Applied onto a pad sound, PWM can be a great way of adding a slash of warmth, somewhat reminiscent to the sound of chorusing.

A triangle wave creates the purest sound of the four types, containing just enough harmonic material to make it still worth putting through a filter. Triangle waves can be a great choice for a suboscillator, tuned an octave down from the main oscillator. Try this on a sawtooth bass, creating a woofer-shaking subsonic to the patch.

The Filters (B)

Once the waveshapes have been selected and mixed, they are passed on to the circular filter section. As previously mentioned, the filters subtract harmonics from the original waveshapes – making the sound darker or thinner depending on the exact filter settings. On a musical level, the filters are important for creating the "juicy" sound of analog synthesis, as best exemplified in sounds like the classic TB-303 bassline used in dance music. However, what's unique about the ES2's filter section is the fact that it contains two completely independent filters, opening up any number of different effects only possible with fully modular synthesizers.

Amplifier and Effects (C)

The amplifier section – or in other words, the volume knob – controls the level of sound leaving the ES2 outputs. This section also includes a distortion unit (great for producing dirty, grungy, synth sounds) and a simple chorus/flanger/phaser unit for warming up any lifeless pad sounds (for example, this was a popular trick used on many of the older Roland synthesizers).

Modulation Matrix (D)

Left to their own devices, the Oscillators and Filter won't produce much interest to the sound by virtue of the fact that they are static – in other words, the sound doesn't change with time. By adding in modulation, however, we can start to add interest, dynamics, and movement into our patches. One simple example of modulation could be the process of linking keyboard velocity to the cutoff point of the filter – now, the harder you play, the brighter (or less filtered) the sound will appear. Other principal modulation sources include envelopes, which shape the ADSR characteristics of a sound over time, and LFOs, which can be used to add effects like vibrato or wah–wah.

As you'd expect, the ES2 features one of the most comprehensive and flexible modulation matrices going, with access to almost every facet of the ES2's programming architecture.

LFOs and Envelopes (E)

The bottom section of the interface contains the controls for the three main modulation sources, including two LFOs and three envelope generators. The key concept to grasp here is that each LFO and envelope has a slightly different design, allowing each to excel in different tasks. LFO 2, for example, can be synched to the tempo of the song, while Envelope 1 provides quick and easy access to AD (Attack Decay) envelopes, which are particularly useful for percussive sounds.

7.4   Working with Oscillators

With an understanding of the overriding synthesis architecture of the ES2 under our belts, let's take a closer look at creating sounds on the ES2.

The first step in synthesizing your own sound on the ES2 should be to consider how you intend to deploy the various oscillators – do you, for example, just use a single oscillator, or use a number of oscillators richly detuned to create a thicker texture? Deactivating the oscillators can be quickly achieved by clicking on the numerical legend to the side of each waveshape pot, this will also have a corresponding reduction in the amount of CPU processing used, an important consideration for those using G4 processors.

As with many parts of the ES2, each oscillator can be seen as having some unique attributes, as well as features common to all three oscillators. Looking first at Oscillator 1, you should see a relatively standard set of synthesizer waveshapes, including sawtooth, pulse, square, triangle, and sine waves. One of the more intriguing options includes the wavetable at the

Figure 7.7 In this example, we've used just a single oscillator to create a simple synth bass sound.

bottom of the dial (preset to a sine wave), and a fully variable FM option (for more information on this, see Section 7.7). To audition the various waveshapes available as part of the wavetable, try clicking on the name and dragging the mouse up or down to scroll through the options.

Oscillators 2 and 3 add furthermore interest with the addition of a fully variable pulse width – morphing between a hollow square wave at the top of the dial and an increasingly thin and reedy pulse as you move the pot clockwise. Another important addition is the provision of two synchronizable oscillators, using either a sawtooth or square waveshape, controlled from the "master" oscillator – Oscillator 1. Going further still, Oscillator 2 also includes the option to ring modulate Oscillators 1 and 2, with Oscillator 2 carrying the ring – modulated output. Oscillators 3's "special feature" is the provision of a noise waveshape – a great source for the random harmonic information so important to percussion sounds.

Figure 7.8 Oscillators 2 and 3 include a number of interesting waveshape options – from fully variable pulse width to ring modulation.

In situations where you're using more than one oscillator, you'll want to experiment with their respective tunings. One simple technique is to apply a slight detune, say ±4 cents, using the fine control on the left of the tuning dials. This should fatten up the sound you're producing – the equivalent of three violinists, for example, playing the same musical motif. The use of semitone shifts adds furthermore interest and power to the patch. Try tuning an oscillator (set to triangle or square wave) down an octave (±12 semitones) from the main oscillators to create a subharmonic, or using third and fifth intervals (±3 or ±4 for the third, ±7 for the fifth) to create chord-like effects from a single note.

Beyond the basic detuning of the oscillators, there are a number of other features on the ES2 that can add width and warmth to the sound you're working with. The Analog control, for example, emulates the pseudo-random pitch and filter-cutoff drifts that occur within the imprecise workings of an analog synthesizer. Therefore, by increasing the Analog parameter, you can add an increasing amount of pitch and cutoff frequency drift, effectively making the patch appear warmer and less sterile.

Activating the Unison mode, in either of the ES2's monophonic modes (mono or legato), takes the "fatness" of your ES2 patches to a whole new level. As with the Unison control on a traditional analog synth, ES2's Unison stacks further virtual oscillators (on top of the three usually available in a patch) to create a thick, detuned, chorus-like effect. Going back to the string analogy, Unison is the equivalent of going from three violin players to a whole section, creating a sound that demands your attention in a mix. The actual width of the effect is defined in two parameters – the number of voices assigned

KNOWLEDGEBASE 2

Oscillator Sync

Let's get this clear from the start: Oscillator Sync has nothing to do with synchronizing the oscillators to the tempo of the track, although of course, you can achieve this with the ES2's LFO 2. Truly, Oscillator Sync is "synchronizing" the wavecycles of Oscillator 2, with respect to the pitch Oscillator 1 – in other words, the pitch of Oscillator 2 is controlled by Oscillator 1. At first, this doesn't seem that exciting – simply controlling the pitch of one oscillator from another – however, by freeing Oscillator 2's tuning controls from conventional tuning applications, we open the lid on a whole new world of exciting effects.

Figure 7.9
With Oscillator Sync engaged, notice how the lower waveform is distorted by the wavecycle of Oscillator 1.

To fully understand Oscillator Sync, we clock what exactly happens to the pitch and timbre of Oscillator 2 once it is placed in-sync with Oscillator 1. The effect is most clearly audible when the initial pitch of Oscillator 2 is set higher than Oscillator 1, so try initially tuning the Oscillator up 12 or more semitones. When the oscillator is placed in sync, you'll notice that its pitch return to that of Oscillator 1, although its timbre has become brighter and harsher – almost distorted, in fact. Looked at under the microscope of an oscilloscope (try rendering the synths output as audio file and taking a closer peek in the Sample Editor), you should see the waveshape being forcibly retriggered, even though it begins its oscillation at a higher pitch. Therefore, it is this periodic retriggering that we perceive as the final pitch of the oscillator.

Oscillator Sync forms one of the most fascinating ways of distorting a waveshape, essentially creating a different version of the waveshape based on the relative pitch difference between the two oscillators. Although it can be used as a static effect, Oscillator Sync works best when the pitch of Oscillator 2 is modulated in some way (try using Envelope 1 for a classic "sync sweep" lead sound), creating dramatic timbral sweeps even before you've reached the filter!

Figure 7.10 Oscillator 1 includes a full wavetable made up of a number of digital waveshapes. Try Ctrl-clicking on the waveshape name to access the list as a menu.

Figure 7.11 Activating the Unison control will layer several "virtual" oscillators on top of the three you're already been using in the patch. This creates a thicker and warmer output to the ES2.

to the unison and the Analog parameter discussed earlier. In Unison mode, therefore, Analog changes its role somewhat and simply sets the amount of detune between the "doubling" oscillators.

Unison, of course, is a staple feature of many of the expansive lead synth sounds found in trance music; although used in moderation, it can also be great way of creating some super-phat basslines. But, like all good things, Unison comes at a cost – namely, your CPU resources!

7.5 Filters, Amplifiers, and Modulators

With the initial pitch and timbre defined by our oscillators, then comes the turn of the filters and amplifiers to shape the raw sound into something that grows and develops over time. The Filter section – found in the large central dial of the ES2 – removes unwanted harmonics from the rich, waveform-only output from the oscillators. For example, starting with a "buzzy" collection of sawtooth waveshapes, you could apply a strong low-pass filter (LPF) to remove the high-end harmonics, leaving you with a soft and warm pad sound. Alternatively, use any one of the other filter types and filter-routing combinations – like band-pass (BP), high-pass, or peak filter types – and you can achieve a real wealth of different timbres. Also, it's worth noting that the ES2 actually contains two independent filters – either allowing you to layer two different filter settings, or use two filters one after the other.

KNOWLEDGEBASE 3
Filter Types
A large part of the "sound" we associate with a synthesizer is brought about by the performance of its filter. Filters come in many shapes and sizes, or more specifically, types and strengths, each with their own characteristics and applications.

The first thing to consider is the strength of the filter – which in the ES2's case is measured in three different decibel levels: 12, 24, and 36 dB. As you'd expect, the higher the decibel rating, the stronger

(Continued)

the filter's ability to attenuate frequencies above or below the cutoff point. Interestingly though, a stronger filter isn't always the most desirable – most of the early Roland synths, for example, took a large part of their charm from the relatively "weak" 12 dB filters. A Moog, however, was famed for its steeper, 24 dB filter. On the whole, I tend to use the stronger filters on darker patches, whereas the classic, 12 dB filters work wonders on TB-303-inspired patches with lots of squelchy resonance.

The type of filter (Lo, Hi, Peak, band reject [BR], band-pass [BP]) defines the form of frequency attenuation taking place. The principal filter type is arguably a LPF, which attenuates frequencies above the frequency cutoff point – transforming the initially bright output of the oscillators into a more rounded overall tone. In most musical applications, the LPF makes most sense, and there are strong similarities to what happens within the acoustic instruments. For example, a piano note will start initially bright (filter open) with high-frequency energy decaying as the note decays (filter gradually closes). A high-pass filter (HPF), however, works in reverse so that frequencies below the cutoff point are attenuated. HPFs, therefore, are a great way of thinning sounds out, especially if the mix is starting to sound heavy and cluttered.

Band-pass and band-reject filters (labeled BP and BR on the ES2) are combination of LPFs and HPFs used to selectively modify a small band of frequencies. BP, for example, allows only a small band of frequencies to pass through, producing a characteristically small-sounding output. BR, however, leaves most of the audio bandwidth untouched and instead attenuates just one "small" band of frequencies, much the same as a tightly Q'd dip on a parametric Equalizer. Conversely, a peaking filter applies a sharp boost at the given cutoff point, again similar to the EQ comparison, but this time a boost rather than a cut.

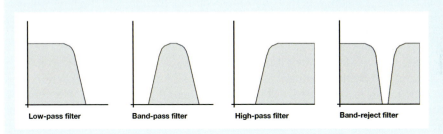

Low-pass filter   Band-pass filter   High-pass filter   Band-reject filter

Figure 7.12
The three principle filter types of subtractive synthesis: Low-pass, Band-pass, High-pass and Band-reject.

Cutoff and Resonance

The two main parameters with both the filters are Cutoff and Resonance. Cutoff, as the name suggests, is the point where the filter begins attenuating frequencies – so, in the case of a LPF, a lowering of the filter will move the cutoff point slowly down the audio spectrum, producing an increasingly darker output. In effect, therefore, you're letting the low frequencies "pass," while the higher ones are attenuated.

Resonance, however, creates a small boost around the point of the frequency cutoff, arguably making the filter's attenuation more pronounced. The exact effect of resonance, though, can vary both between the filter types and the relative position of the cutoff. Using the example of a LPF, a high-cutoff setting with

Figure 7.13
The filter section shapes the timbre of the raw output from the ES2's oscillators section. The two main parameters to interact with are filter Cutoff and Resonance.

some resonance applied will produce an increasing thin output (as the resonance boosts frequencies at the top end of the audio spectrum). Alternatively, resonance applied on a LPF down the bottom of its scaling will produce an output with increasing amounts of low-end – an effect that can often be useful on bass sounds. Take the resonance all the way to its maximum setting, though, and the filter will self-oscillate (in other words, produce a high-pitched ringing sound almost like a sine wave) at a frequency defined by the filter-cutoff position.

Ultimately, the point to note with the filter is that it provides a large part of the "character" in a synth sound, probably best exemplified by the dripping, acidic resonance of a TB-303 filter in action on a bassline. All sounds, to a greater or lesser degree, will use the filter, and if you can add some dynamic movement or expression into the patch (maybe responding to velocity, for example, using an envelope and/or LFO), this character will become even more pronounced and interesting to the ear.

Changing Volume Over Time

By default, the amplifier is controlled by Envelope 3, with the two remaining envelopes available for use with any other application you see fit. Envelopes allow you to manipulate the amplitude (or any other parameter, for that matter) over time – raising and lowering the given level for the duration for which the key is held down. The progression over time is defined by a number of key parameters: attack, decay, sustain, and release. Attack, like the majority of the other ADSR parameters, is a time measurement, so that a low setting produces a quick attack (useful for sounds that are more percussive in nature), whereas a high setting produces a graduated attack that, for example, might be applied to pads. Decay defines the time taken for the note to fall down to the sustain level, while release defines the final drop to silence after the key is released.

Looking at any sound in the natural world, it's easy to see how we can begin to define their qualities by the characteristics of an amplifier envelope. A piano, for example, would have a quick attack, slow decay, no sustain, and a relatively quick release. However, a smooth, evolving string sound might have a slow attack, slow release, medium-to-high sustain, and a slow release. So, coupled with the basic waveform selection and the attenuation of the filter, we can really start to establish the defining properties and uniqueness of the patch we're creating.

Figure 7.14 By default, Envelope 3 is patched to the amplifier, allowing you to shape the volume of the patch over time.

Of course, alongside Envelope 3, the ES2 also contains two other envelopes, but to understand how these are used we need to look at the modulation matrix.

Bringing It All Together: The Modulation Matrix

Tying all the elements together is the modulation matrix, which links a number of modulation sources in the ES2 (like the aforementioned Envelopes, LFOs, external MIDI controllers, and so on) to destinations within the synthesis architecture. The complexity and depth of the modulation matrix is what sets the ES2 apart even from many dedicated third-party synthesizers and provides an almost unparalleled amount of flexibility and creative potential for synthesis. The only catch, however, is that certain elements of the ES2's control panel often need to be connected through the matrix before they have any effect on the sound!

Looking more closely at the matrix, we can see a number of different routing paths (ten in total) indicated by the line of blue boxes. Each routing connection will have a target parameter (which could be a parameter like Filter Cutoff or Pan), a modulation source, and the amount of modulation that can be applied. Just a quick glance through the respective menus for both the target and source reveals the depth of the ES2, with over 40 program targets and over 20 possible modulation sources, with everything from standard modulation sources, like LFOs and envelopes, to random generators (RndNO1 and RndNO2) and external MIDI controllers.

Some of the principal connections you might want to explore involve controlling the filter from modulators like Envelopes 1 and 2, and the two LFOs. The LFO provides a "periodic" form of modulation – in other words, a repeating continual shape or pattern that modulates the parameter, rather than an

Figure 7.15 The ES2 modulation matrix allows us to map up to ten modulation sources (LFO, envelopes, and so on) to different destinations (Filter Cutoff, and so on) within the synthesizer's architecture.

ADSR-type modulation with fixed stages. Arguably the best example of this is vibrato – where an LFO would be routed to the Pitch123 target. However, LFOs can also be an excellent source of Tremolo-type effects (routed to Amp) or a wobbling wah–wah (routed to Cut 1 ± 2).

Figure 7.16 Each of the two LFOs can be set to a variety of different modulation shapes, each creating a unique form of modulation effect. For example, the sawtooth waveshape, or the stepped sample-and-hold waveshape can be great for sequencer-like effects, especially attached the filter. In that respect, it's worth noting that, in its below "DC" setting, LFO 2 can also be set to sync with the project's tempo.

KNOWLEDGEBASE 4
Vector Synthesis

Vector synthesis, originally developed as part of the Prophet VS synthesizer, was first envisaged as a unique system for adding movement into a patch. In the case of the Prophet VS, it used a series of four oscillators, or sources, labeled A, B, C, and D, which could be morphed, using a vector joystick control. By storing this movement as part of a vector envelope, the patch could then recall this dynamic movement creating timbral movement impossible to achieve elsewhere.

By switching the modulation matrix from Router to Vector, you can activate the ES2's own vector-synthesis features. The vector controls themselves work with both the triangular oscillator vector mixer (using three oscillators, rather than the Prophet VS's four), but they can also make use of an additional vector controller in the form of the XY pad. By default, the XY pad has no assignment, but by using the X Target and Y Target parameters, you could, for example, assign the X- and Y-axis to Cutoff and Resonance – providing complete gestural control over two parameters on the one pad! To create a basic vector envelope, click on any of the vector nodes and position both the oscillator mixer and the assigned XY pad. Now, select the other two nodes in turn, ensuring each has a different position for the oscillator mixer and XY pad.

To hear the vector envelope in action you'll need to uncheck the Solo Point switch – you should now hear the ES2 morphing between its various states. To add further nodes into the envelope, simply Shift-click on the time line. You can also set the time between any of the nodes by click-dragging on the various ms settings, making certain transitions quicker than others. Changing the loop mode will then allow the vector envelope to loop, based on the direction set and the relative position of the two markers – a sustain marker (where the envelope would hold, if no loop were set) back to the loop marker, indicating the full duration of the loop.

(Continued)

By exploring some of its deeper features, you can also turn the vector envelope into a surprisingly effective source of rhythmic pads and textures. Try changing the Loop Rate from "as set" to a tempo division like 1/2. Now, the vector envelope appears to pulse in time with the track, a feature that could really be exploited with further nodes (up to 15 are allowed) and some interesting tuning and waveshape assignments of the respective oscillators. If you're in any doubt as to the effectiveness of this, try listening to some of the Vector Rhythm (Curve) presets as part of ES2 Legacy folder.

Figure 7.17 The ES2's vector envelope can add whole new levels of movement into your patch, by directly manipulating the oscillator mixer or XY pad.

7.6 Global Parameters and Output Effects

As part of the ES2 global parameters to the right of the filter dial we find a number of useful tools to add further layers of interest to the sound. The Sine Level, for example, adds an additional sine wave element to sound based on the pitch of Oscillator 1. This is a useful tool for general "bass fattening," especially if the output of your filter has left your patch sounding slightly weak. If you want more grit, though, the inbuilt distortion unit, which is switchable from hard to soft operation, helps in giving aggressive synth sounds considerably more edge. Even in relatively soft settings, with small amounts blend in, the distortion can make a patch sound slightly more analog than its digital origins would suggest!

Figure 7.18 The output effects are a useful means of adding extra warmth (using the chorus unit) or grunge (using the distortion) to the ES2 patches.

KNOWLEDGEBASE 5
ES2 Filter Routing

One of the most intriguing aspects of the ES2 is its immensely flexible filter-routing system. Essentially, the two filters can be connected either in series – so that sound flows sequentially from one to the other – or in parallel, where the sound is split and fed to each filter, individually. To switch between the two filter modes, simply click on the filter's Parallel or Series switch accordingly, and enjoy the sight of the revolving filter controls! Taking this up to another level, we then have the Blend parameter, alongside the Drive control, both adding a whole new level of interaction based on the current filter arrangement and its relative position.

Looking first at the Parallel routing (where sound is split and sent through to two filters individually), the Blend parameter can be seen as a simple crossfade parameter, making one filter's output more prominent than the other in the final mix. In serial mode, the concept is a little tricky to understand, although not impossible. In its central position, Blend creates a true serial configuration with Oscillator 1's output being passed to the Oscillator 2's. Either side of the line, however, an amount of either Filter 1 or 2 is progressively bypassed, effectively creating a bias toward one of the two filters. Of course, at either extremes of the Blend parameter, the output is exclusively Filter 1 or 2, respectively.

The Drive parameter adds polyphonic distortion either before or during the filter's operation (again, based on the filter's routing). Polyphonic drive differs from the conventional distortion (available in the amplifier section) in that it distorts each voice individually – in effect, using a different distortion module for each note played! In short, this means that a chord can be given a little "grit," rather than collapsing in intermodulate distortion, as it would in a conventional single distortion module setting. In Parallel routings, the drive is placed before the two filters, effectively, giving them slightly more harmonic material to play with. In Serial configurations, the Drive is generally placed after Filter 1 (although the exact interaction with Filter 2 varies given the Blend parameter), making the distortion effect more pronounced, especially on high resonance settings.

Figure 7.19 Here are two ES2 synthesizers, each set to one of the two different filter routings – parallel or serial. Using the Blend and Drive parameters adds further possibilities to how you can use and abuse the ES2's filters.

The Modulation Effects section provides inbuilt chorus, flanging, and phasing effects, with a variable intensity and speed. All of these are useful warming effects, adding an extra sense of depth, stereo width, and movement to your ES2 patch that really seems to suit pad-like sounds.

7.7   Retro Synth

If you've conquered the depths of the ES2, then the relative simplicity of the Retro Synth will make it a breeze to understand. In essence, the Retro Synth is designed with immediacy in mind, with a simplified and direct set of user controls, much like many of the Roland synthesizers from the early 80s. While it lacks some of the depths of the ES2 (especially with respect to modulation routings), it can certainly produce some powerful, sonically effective sounds. Equally, the "Retro" concept is more than a gimmick, with a four distinct synthesizer engines that make it closer to four instruments than one!

Figure 7.20 Retro Synth comes in four different flavors, thanks to a switchable "synthesis engine" parameter.

7.8 Synthesis Engines

The key concept to grasp with Retro Synth is its four distinct synthesis engines, each of which mimics a style of synthesis from the 70s and 80s. Rather than changing the entirety of the instrument, though, the majority of the controls – namely, the filters, envelopes, and LFOs – are common to all the engines. As such, you can move between the different engines and still find a large part of your sound "translating" to the alternative setting. Equally, there're some interesting serendipitous results that can come from randomly moving between the engines, even from just working within the supplied presets.

Analog Mode

The Analog mode is designed to mimic the traditional subtractive synthesizers from the 70s and 80s. The key components are two oscillators, which have a variety of subtractive waveforms including sawtooth, triangle, a fully-variable pulse wave, and noise. The controls in the oscillator section are relatively straightforward, therefore, with a Vibrato control, Semitone and Cent detuning, and a Mix control. With respect to the pulse wave, you can also use the Shape Modulation control to modulate pulse width using either the LFO or Filter Env.

Sync Mode

The Sync mode is almost identical to the Analog mode, although in this example there's a hardwired oscillator sync control (which we covered in Knowledgebase 2). Notice that the shape modulation changes to a Sync Modulation control, which is most effective when it's set to Filter Envelope if you want to replicate the classic Sync-sweep sounds from the 70s and 80s.

Figure 7.21 The Analog mode models a classic subtractive synth from the 70s or 80s.

Figure 7.22 Sync Mode is almost identical to Analog mode with the additional of a hard-wired waveform sync control.

Table Mode

Table Mode is based on vector synthesis (detailed in Knowledgebase 4). The key point to note here is that we have a fully variable shape control for each oscillator, with a range of digital waves stored in the wavetable. In Table Mode the modulation control is set to modulate the choice of waveform Shape, creating the distinctive "wavetable sweep" sounds heard from the PPG Wave 2.3. Although the extreme sweeping effects are interesting, the best results seem to come from a subtle application of the LFO so that the Shape modulates in a slow and graduated fashion.

Figure 7.23 Wavetable mode models the classic PPG synthesizers from the mid 80s.

FM Mode

FM Mode is easily the trickiest synthesis engine to understand, by no less immediate if you just want to start tweaking a few controls and having fun. In this arrangement, Retro Synth works as a simple two-operator FM Synthesizer, with one oscillator working as Modulator, the other as Carrier (hear the difference between these two by moving the Mix Control). In essence, adjustments to the Harmonic, Inharmonic, and Shape controls adjust the tuning and shape of the Modulator. Changing the about of FM sets the degree of "frequency modulation" applied by the Modulator to the Carrier. Sounds easy, right? Well, if in doubt, just fiddle with the controls until you get something that sounds interesting to your ears!

Figure 7.24 The FM Mode sets up an arrangement between a Carrier and Modulator, much like the EFM1.

7.9 Retro Synth Common Controls

The remaining controls on Retro Synth are common to all Synthesis engines. As a result, a large proportion of your sound's shape and tone will be transferred even as you move between radically different synthesis engines.

Filter

The filter section contains eight different modes (low-pass, high-pass, and so on) with variable strengths, including 6 dB, 12 dB, and 24 dB operation. Once you've established the cutoff and resonance, the main controls relate to modulation. Use the Key parameter so that the filter tracks your keyboard position, while LFO and Env set the amount of modulation applied by the LFO and Filter Envelope, respectively.

Amp

The amplifier section is rather basic, with just an overall Volume control and Sine Level. The Sine Level option is useful in situations where large amounts of resonance or high-pass filtering have removed the fundamental frequency. The Sine Wave is at the pitch of Oscillator One and avoids the filter by being routed directly to Retro Synth's output.

Effect

The effect section contains a simple Flanger and Chorus effect. Both of these controls were important means of adding warmth and depth to classic analog synthesizers, and are an almost essential part of a synth pad sound.

LFO/Vibrato

The bottom half of the Retro Synth interface deals with the various sources of modulation, including individual LFO and Vibrato sections, as well as the two envelope generators. The LFO and Vibrato controls are almost identical, with Vibrato being hardwired to oscillator pitch, while the LFO can be routed to a number of destination including Shape

Figure 7.25 The remaining controls of the Retro Synth are common to all of the different synthesis engines.

Modulation, and the filter. Both the LFO and Vibrato sections feature a variety of different modulation waveshape, variable rate, and the ability to tempo-lock the modulation using a Sync control.

Filter and Amp Env

Both the Filter and Amplifier Envelope are hard-wired to their particular destination, with simple ADSR controls governing the shape of the modulation. The only important addition control to note is the Vel

Figure 7.26 The lower part of Retro Synth's interface covers all the important modulation sources, including LFOs and envelopes.

control, which is used to establish the amount of Velocity control applied to the depth of modulation. In the case of the Filter, harder velocities could make the sound bright, while on the amplifier velocity could be used to make the sound louder or quieter depending on how hard the notes are played.

7.10 Retro Synth's Setting Page

The setting page is worth exploring if you want access to some of Retro Synth's deeper levels of functionality. The Double control, for example, adds instant depth and dimension to the sound, especially if Stereo Spread is set to its full setting and a small amount of Voice Detune is being applied. By default, Double is active, so it's more the case that you might want to deactivate the feature if you want a slightly smaller sound.

The Controller section is also useful if you're intending to control the Retro Synth using your modulation wheel, for example, or explore some more creative opportunities in respect to velocity control. The modulation sources are rather limited, as are the four possible destinations, but you do get the possibility of exploring a greater amount of modulation options.

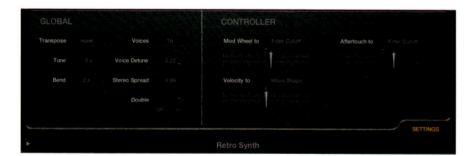

Figure 7.27
Access the Settings page to explore some of the deeper functions behind Retro Synth.

7.11 EFM1 and Frequency Modulation Synthesis

Subtractive synthesis and the features of the ES2 might cover a large part of the sounds used in contemporary music production, but there are other methods of synthesis – like frequency modulation (FM) – with plenty to offer. EFM1 is Logic's solution for FM synthesis (although FM also makes some honorable appearance in the ES2) using the same principles carved out in Yamaha's infamous DX range of synthesizers back in the 1980s. Although the EFM1 lacks some of the depth and complexity of a full-blown seven operator DX7, or Native Instruments' FM7, it remains a valuable insight into a unique form of sound creation and a valuable contrast to the sounds produced by the ES2 and Logic's other subtractive synthesizers.

Not surprisingly, the heart of FM synthesis, and the sounds produced by the EFM1, center on the process of FM – but what is it, and how can it be used to produce musically useful results? The process of FM requires two oscillators – one oscillator acting as a Carrier, and the other oscillator acting as a Modulator.

You can see this clearly on the EFM1 front panel, with the carrier on the right-hand side of the panel, and the modulator to the left-hand side. The relationship between the two is that the modulator is modulating the pitch, or frequency, of the carrier, with the carrier oscillator being the only oscillator present in the audio output.

In many ways, the simple modulator/carrier configuration we see as a basis of the EFM1 bears a fundamental similarity to an LFO modulating the frequency of an oscillator on the ES2. As the LFO is working below the audio spectrum (around about 6 Hz), we hear the effect of this modulation as vibrato, with stronger amounts of modulation creating an increased amount of pitch wobble.

However, in the case of EFM, the modulator is working right up in the audio spectrum, producing an effect quite unlike vibrato. By modulating pitch using an oscillator in the audio spectrum, we effectively distort the original waveshape, creating new harmonic components – the stronger the amount of modulation (set with the large central FM parameter), the greater the amount of the distortion and hence the richer the eventual harmonic output.

One of the most important relationships to understand in FM synthesis is the ratio between the pitch of the carrier and the modulator. On the EFM1, the tuning of the modulator and carrier is established using the Harmonic control on either side of the interface. Try setting the carrier to the first harmonic and experiment with the modulator on first, second, and third harmonics. With both carrier and modulator set on the first harmonic, the sound will approximate a sawtooth, with a collection of both odd and even harmonics.

Figure 7.28 By increasing the amount of FM, we create a more harmonically rich output in much the same way as opening a filter on a conventional subtractive synth.

Figure 7.29 Changing the relative tuning of the modulator and carrier changes the resultant harmonics produced as the FM knob is increased.

Moving the modulator up to the second harmonic will produce a more hollow and open timbre, with an output largely made up of odd harmonics. Going further still, this time up to the third harmonic, the sound becomes more bell-like, by the virtue of the fact that you're using an odd harmonic relationship.

Modulation on the EFM1

Okay, we've established the basics of a "static" FM sound – namely, the tuning ratios between the carrier and modulator, and FM depth – but how do we go about adding movement or interest into the patch? Let's first consider trying to create a basic filter sweep. On an analog synthesizer, this would be achieved using an envelope patched to modulate filter cutoff, with the envelope opening or closing the filter over time. In the EFM1's case, we can use the Modulation Envelope toward the top of the interface in conjunction with the FM Depth parameter to the right of it. By setting up an appropriate envelope (quick attack, graduated decay, no sustain) with FM depth set to its maximum setting and the large FM dial set to its center position, the EFM1 should then gradually reduce the amount of FM as the key is held down.

Figure 7.30 Use the Modulation Envelope and FM Depth parameter to modulate the large FM dial over time. The effect of this modulation would approximate a filter sweep on a traditional analog synthesizer.

Another form of modulation is the LFO, which, in the case of the EFM1, can be moved between being applied to pitch – to create vibrato – and being routed to the FM depth. As we've seen in the last example, routing a modulator to FM depth produces an effect comparable with filter modulation. In this case, therefore, routing the LFO to FM creates a form of wah–wah effect, with the timbre of the EFM1's output wobbling over time. An entirely more chaotic form or modulation, though, is possible by routing the modulation envelope to the modulator pitch (using the dedicated pot). The effect is an extreme, and possibly dissonant, change in timbre, which can be greatly used in aggressive synth sweeps.

Refining Your EFM1 Patches

To further refine your patch, the EFM1 includes a number of different parameters, which don't differ too greatly from the features found on the ES2. The Volume Envelope, for example, allows you to shape the amplitude over time – for example, try creating bell sounds using a quick attack, no sustain, and a slow decay and release. The suboscillator, as on the ES2, doubles the synthesizer's output with a suboscillator an octave below, giving bass sounds a real punch. Finally, the Stereo Detune option effectively creates a complete double of your patch, slightly detuned and with a distribution (of the two engines) across the stereo image. Used subtly, this can be a simple solution for adding warmth, depth, and width to any of your EFM1 patches.

KNOWLEDGEBASE 6

Wavetable Synthesis

Wolfgang Palm originally developed the Wavetable synthesis in the early 1980s as a part of his groundbreaking work on the PPG wave synthesizer. The principle was that an "audio event" – like a filter opening – could be broken down and stored in a Wavetable constructed from 64 single-cycle waves. By stepping through these wave cycles, and interpolating between the various settings, the PPG could create unique and interesting textures, often characterized by a glassy, shimmering tone. Although PPG eventually went bust in the late 1980s, Palm's intriguing Wavetable system continued to live on (at least for a few more years!) courtesy of Waldorf's Microwave and Microwave II synthesizers.

By including its own digital Wavetable, the ES2 pays homage to this late great synthesizer, and, thanks to its flexible modulation matrix, can recreate many of the sounds that the PPG wave was so famous for. If you want to experiment with Wavetable synthesis, the trick is to select OscWaves as a modulation target and ensure the oscillators are each set to their Wavetable position. Choosing the exact type of modulation source (albeit an LFO, envelope, the modulation wheel, or something like keyboard velocity) will then dictate the type of movement and the sound produced. Try using a slow LFO (0.10 Hz, for example) and gently modulate the Wavetable portion to create a shimmering digital pad. Alternatively, you can create an interesting glitch-like effect by modulating the Wavetable using a fast AD envelope.

To add further interest into the equation, you can also modulate each of the Wavetables individually, possibly using a different modulation source (LFO 1, LFO 2, and modwheel) on each. To do this, you'll need to use OSC1Wave, OSC2Wave, and OSC3Wave destinations, respectively.

Figure 7.31 By modulating the Wavetables within the ES2, you can create some interesting "moving" synth sounds that are reminiscent of the PPG Wave 2.3.

7.12   Component Modeling: Sculpture

When it comes to cutting-edge sound design, Sculpture represents one of the most exciting and perplexing parts of the Logic, and also a real indicator of a future development of synthesis, and a component that may well become increasingly dominant in sound production as a whole. Sculpture's greatest strength – but equally the reason why it takes so much time to master – is that it throws out the conventional "rule book" of synthesis. You won't find traditional features like oscillators or amplifiers (although filters do get a minor input!), but instead parameters like Media Loss and Exciter Objects, producing sounds that are almost impossible to achieve elsewhere!

Like an increasing number of virtual instrument – including Arturia's Brass and Applied Acoustic Systems' String Studio – Sculpture is built on the concept of physical modeling, or, more specifically, component modeling. The basis of all these instruments is a complex mathematical model of an instrument, with a precise understanding of how sonic variables (the materials the instrument is constructed from, and so on) and various interactions (in other words, how it is played) affect the sound that the instrument produces. In the case of Arturia's brass, for example, the "model" is based on brass instruments like trumpet and trombone, whereas in the example of Sculpture, we have the multitude of possible musical outcomes that result from interactions with a string!

7.13   Objects

The starting point of patch in Sculpture is to consider how your instrument is constructed, and the means in which it will be played. Indeed, in the first part of designing your sound, you will spend a lot of time moving between two key areas of Sculpture's interface: the Material Pad, in the center on the control panel, and three different Exciter Objects, all interacting with each other and having some fundamental impact on the type of sound being produced.

Let's first look at the concept of Objects. To produce any note from a real instrument we need to introduce energy into it – for example, blowing across the reed of a clarinet or using a stick to hit a snare drum. Correspondingly, in Sculpture, we have a choice of up to three Objects that can be introduced to the string to begin creating a note: it could be hit, pluck, struck, or blown (to name but a few of our potential objects), with each different object producing remarkably different results. What this immediately means is that you need to have at least one object active to create any audio output whatsoever. Also, by having more than one object, you can create some weird-and-wonderful effects like a plucked string being gently dampened half way up its length, or complete audio chaos with a number of objects all playing their part at the same time! As you can see, the best results will come with a considered application of these settings, rather than just throwing it all "into the pot."

Figure 7.32 Sculpture's string can be played in a number of different ways, defined by the Type parameter of each object. For example, Object 1 includes picking, bowing, and a number of other "interactions."

Accompanying each exciter object, you'll find another three key parameters – Strength, Timbre, and Variation – that set some of the initial effects of the object against the string. Of course, with each different object model the precise impact of the three parameters varies, so it is worth spending

Figure 7.33 Objects 2 and 3 contain a number of different string interactions, which can be used to enhance the basic settings established with Object 1 – like the string being picked and damped at the same time.

some time with each to understand the range of sounds and effects you can start to produce from them. If you've got the processing power, it might also be worth turning on the string animation by Ctrl-clicking on the thin green line toward the middle of the object controls.

Although you can produce some fantastic sounds with just the one object, the true potential and interest with Sculpture comes as you add further objects into the equation. In particular, you'll notice that Objects 2 and 3 contain some of the more esoteric object options, like Disturb 2-Sided, which approximates the effect of a ring placed around the string, or Mass, which places a large weight at some point along the string. Note also that you can set the object to be triggered in one of three ways – KeyOn, Always, and KeyOff – so that an object is only triggered after the key is released.

If you do use more than one object, it's worth paying close attention to the relative strength of the various objects. For example, using all three objects at maximum strength tend to create an overwhelming sound with little or no character. However, picking one key object (usually using Object 1) and augmenting this with subtle additions of other objects seems to produce the more interesting hybrid effects. For example, you could choose to use a Strike model on Object 1, which is then lightly Disturbed (the effect of placing an object on the string) using Object 2.

As important as the types of object used to initiate vibrations in the string, we also need to consider where along the string it is played. Take, for example, a nylon string guitar being played using a small pick. Played partway down the length of the string, just above the sound hole, results in a pleasing, rounded tone. However, if we were to play right up against the bridge, the tone will change dramatically, becoming harsher, with a thinner overall timbre. Just as with real life, therefore, we can position each of the three objects along our virtual string using the Object Positioning display in the center of the Objects section.

Again, the best way of understanding how the models work, and their precise effect at each point along the string, is to experiment, although the general principle follows that a fuller sound emanates from the center of string, whereas the extremities tend to produce a more nasal timbre. Adding in additional objects makes the variables ever more complex, although again, the general rule-of-thumb is that objects arranged closer together tend have a more pronounced effect on one another.

As part of the same object positioning display, you'll also find controls for each of Sculpture's virtual pickups. These work in much the same way as the electromagnetic pickups on an electric guitar, which of course, tend to output a different sound based on the relative position along the guitar string. In Sculpture's case, we're presented with two pickups that can be slid along the virtual string, contributing to the output in a number of different ways. To the far right of Sculpture's interface, you'll also find the stereo control, which can define a stereo spread both for Sculpture's key range (moving from left to right as you move up the keyboard) and between the two pickups. Try using a wide stereo with the two pickups so as to create pleasing width to Sculpture's output – an effect especially useful on pad sounds.

Figure 7.34 The position of an object can have a big effect on the string's vibrations. This interaction becomes even more complex with a greater amount of objects.

Figure 7.35 As with an electric guitar, the restive position of the pickups (A and B) will also influence the sound produced. Use the Spread control on the right-hand side of the interface to turn this into a stereo effect.

7.14   The String

Having spent so much time analyzing the objects used to excite Sculpture's string, we've neglected to look at an equally fundamental part of creating sound in Sculpture – the design of the string itself. Again, relating principles back into real life, we can see that the way a string is made will have a big effect on the sound produced, as evident in the apparent difference between a nylon-strung acoustic guitar and steel-strung acoustic guitar. By adjusting properties in the central Material Pad section, therefore, we can fine-tune the properties of our virtual string – morphing the sound between glassy, bell-like textures to warm, woody marimba tones, and plenty more besides!

The central X/Y controller of the Material pad sets the relative Inner Loss and Stiffness of the virtual string, although the precise qualities of this is better expressed by the small nylon, wood, steel, and glass legends in the four corners of the XY pad. Try moving the small control ball and listening to the harmonic spectra as you move around the pad – for example, listen out for the brighter tones toward the bottom of the XY pad against the move toward more bell-like harmonics as you move from left to right. As with objects placed along the string, this simple pad has one of the biggest impacts on the sounds you produce from Sculpture, so it's worth getting to know the various initial sounds possible from it.

Figure 7.36 Using the central X/Y control will allow you to change the material properties of Sculpture's virtual string.

Figure 7.37 Parameters around the Material pad set a number of other physical considerations. Media Loss, for example, could be thought of as a form of decay or release setting, governing the loss of energy over time.

Around the Material pad, we can also find a number of other key parameters in relation to the design of the string: Resolution, Media Loss, and Tension Mod. The most important of these is the Media Loss parameter, which sets the dampening applied to the string based on its surrounding conditions. A more meaningful way of understanding this would be the "release" of the sound, so that with a lower media loss setting, the sound will appear to decay over a long period of time, whereas quicker media loss settings tend to produce a shorter, more clipped sound.

Tension Mod corresponds to momentary fluctuations in pitch as the string is excited – the more the Tension Mod is applied, the greater these pitch fluctuations appear in the final output. Used subtly, Tension Mod can add a discrete sense of the natural pitch shifts that occur within a real acoustic instrument, although care should always be taken not to overuse features like Tension Mod as it could make the instrument sound like it has poor intonation! Resolution, however, corresponds to the amount of harmonics used to generate Sculpture's output. Try experimenting with lower resolution settings, as this can be an interesting way of creating inharmonic overtones (great for bell sounds) and as a means of reducing Sculpture's CPU load.

Taking the string design up another level, we can also add variations to its design based on the relative key position, otherwise known as Keyscaling, and in its release phase. So, for example, as you play up the keyboard the sound could change from a warm, woody timbre to a bright, glassy string. Equally, Keyscale Media Loss could approximate the shortening of strings duration as you move up a piano keyboard, for example. To modify any of these settings, simply click on the Keyscale and Release bottom at the bottom of the Material Pad, and adjust the parameters (using the colored indicators) as appropriate.

As a closing thought to the Object and String controls, try to remember that experimenting in component modeling synthesis is rather like relearning the art of synthesis. So, rather than using tried and tested means of solving problems – like adjusting an envelope to change the dynamics of a sound, or a filter to change the timbre – you'll need to think "outside the box" and discover a new palette of solutions available in a component-modeling universe. Changing the amplitude envelope, for example, could involve a change or even an addition of an object onto the strings – a short decay, for example, made by physically dampening the string on the key off. Alternatively, adapting the string's Inner Loss properties, or simply less strength on the exciter, could make for a darker sound. Ultimately, it's all about new solutions to old problems.

Figure 7.38 Using the Keyscale and Release control, you can vary the properties of the string both in the release phase of the note and in respect to keyboard position.

KNOWLEDGEBASE 7

Bounce in Place

Rendering an audio instrument as an audio file has many advantages – not only saving valuable CPU resources, but also allowing you to safely export your project to other DAWs (or other users that don't share the same Audio Unit plug-ins), as well as opening up some interesting possibilities with respect to manipulating the resultant audio file.

Arguably the best way of rendering an audio instrument is to use the Bounce in Place feature, which renders one or more regions, complete with effect plug-ins, to a single audio file. To perform the bounce, simply select the desired regions in the Tracks Area, and then Ctrl + click the region to bring out the contextual menu then Bounce > Bounce Region in Place. The accompanying dialog box will allow you the detail of how the bounce is made – including options to bypass and effect plug-ins, as well as the option to Normalize the file. Once you've performed the bounce, Logic will import the new audio file, place it accordingly, and assign it a new track. The source track can then either be left, muted, or deleted as you specify in the Bounce Regions in Place dialog box.

(Continued)

Figure 7.39 Use Bounce in Place feature to render an audio instrument as an audio file.

Once rendered as an audio file, you can now explore some of the possibilities unavailable in "virtual" form. Chief among these is the ability to reverse the region by entering the Sample Editor, choosing the File tab and then selecting Functions > Reverse.

7.15   Waveshapers and Beyond

Theoretically, in a true component modeling synthesizer, the object and string parameters should be all that is required to create your eventual output. Indeed in Sculpture's case, it is highly likely that you'll spend the majority of your time working in these sections defining your sound, with the finished patch simply requiring a touch of EQ, compression, and reverb to set it in the mix. Thoughtfully though, Sculpture also provides a number of additional features, including Waveshapers, Filters, ADSR envelopes, Delay lines, and EQs, to furthermore enhance and modify the effects you can produce within the plug-in. Although these aren't necessarily integral to the concept of component modeling, they are highly useful additions.

Following the Sculpture's internal signal path, we start with the Amplitude Envelope. The important thing to note here is that the amplitude envelope is post the component-modeling engine – its settings have no direct correlation to the sound produced and merely act as a means of modifying what's already there. So, for example, extending the release phase has no effect if the string has stopped vibrating (instead, consider raising the Media Loss), but a slow attack could be a suitable solution for making a bowed string have an even more graduated entry.

The Waveshaper, which follows after the Amplitude Envelope, is one of the most useful additions to Sculpture's synthesis engine, essentially applying a polyphonic distortion effect (much the same as the ES2's filter-distortion effect). In addition to applying a suitable valve-warming effect (try the Soft

Figure 7.40
The amplifier envelope allows you to change the volume characteristic over time, but, remember, this is initially defined by the string and material properties.

Figure 7.41
The Waveshaper adds grit or warmth to your Sculpture patch.

Figure 7.42
Use the filter as an additional form of timbre control, with various different modes to change the relative effect achieved.

Saturation algorithm), the Waveshaper is essential in producing synth-like tones from Sculpture, making the output of the component-modeling engine much closer to the buzzy, harmonically rich output of a subtractive synth's oscillator. Of course, because the distortion is polyphonic (with each voice being distorted on an individual basis) even strong settings don't result in the sound collapsing in piles of unmusical intermodulate distortion.

Working in close conjunction with the Waveshaper is the filter, which includes five different filter modes – Hi-pass, Low-pass, Peak, BP, and Notch. Given the added harmonic material created with the Waveshaper, you can now use the filter to subtract harmonics, producing synth bass sounds, and so on. Also the filter can be a useful additional way of tapering the output from the component-modeling engine – maybe either using a LPF to create a dark bell patch or an HPF to take the low end out of a steel-string guitar patch. Remember, as with a conventional subtractive synth, the filter is also an ideal destination to control using the modulation wheel, an LFO, or an envelope, as we'll see in Section 7.12.

Three final modules provide the last refinements: a relatively simple Stereo Delay line effect (which isn't entirely dissimilar to Logic's Stereo Delay plug-in), a Body Equalizer, and a Limiter. The body equalizer's simplest application is Low Mid Hi Model, used to cut or boost frequencies at 80 Hz, 2.5 kHz, and 12 kHz, respectively. More advanced treatments are possible using one of the supplied instrument models, which are themselves drawn from impulse response recordings of the respective instrument bodies. In effect, what you have access to here is a detailed EQ profile of the characteristics of the instrument in question – it won't, for example, turn a harp sound into a cello simply through the application of the Cello's Impulse Response, but it does at least impart some of the instrument's tonal character.

Figure 7.43
Use the Body
EQ and the
instrument-based
IR to impose
the timbral
characteristics
of an instrument
onto your
Sculpture patch.

Once loaded, the impulse response can be tweaked in a number of ways to best match the patch you're trying to create. Lowering the Intensity, for example, can be a useful way of softening the effect of the impulse response, imparting a subtle modification of tone rather than completely transforming it. Also, try experimenting with the Formant Shift and Formant Stretch parameters, allowing you to more carefully tune in the respective "lumps and bumps" of the impulse response profile with the tonal characteristics of what you're trying to create.

7.16   Modulation and Morphing

As we saw in the ES2, modulation is essential for bringing a patch to life: making it more expressive in response to actions on keyboard, adding in LFOs to add effects like vibrato and tremolo, and using envelopes to shape the sound over time. As you'd expect, Sculpture is no exception to this rule, with a modulation section (toward the bottom of the interface) that rivals, if not exceeds, what is possible on the ES2. More important, the application of modulation on Sculpture has the potential to directly influence the specific properties of the instrument (string construction, object position, and so on) rather than using crude "signal modifiers" (filters, amplifiers, and so on) to approximate the same effect.

The bottom dark gray pane in Sculpture contains all the various controls for changing the way the instrument responds to a performance and behaves over time, including two LFOs, a Jitter generator, vibrato, velocity modulations, controller routing, two envelopes, a Morph pad, and a Morph pad vector envelope – enough to keep anybody happy! Pressing the various tabs allows you to step through the relevant sections of the basic modulators, while the Morph pad and envelopes hide some intriguing "contextual" menus and functions, such as recordable envelopes and patch randomization.

Figure 7.44 The modulation section of the Sculpture adds an additional level of dynamics and movement to your patch.

Figure 7.45 Routing the LFOs to the pickup position can be a great way of creating stereo movement in a patch.

Simple modulation can be accessed through the Vibrato and LFO 1 and 2 tabs. Vibrato permanently connects a spare LFO to the pitch of Sculpture, whereas LFO 1 and 2 can be wired to any number of destinations within the Sculpture patch – from the usual Filter Cutoff, and so on, to some of the more unusual component-modeling parameters like Object Position or Pickup positioning. Exploring modulation with these parameters, in particular, can lead to some of the most musically useful and rewarding applications of Sculpture, with complex or subtle modifications in timbre that would be impossible to achieve on a traditional subtractive synthesizer. For example, try gently modulating the pickup position in a wide stereo pickup spread – the effect is a great way of adding subtle movement to a patch and palpable sense of "stereoness."

For direct response from a MIDI performance, Sculpture features two controller paths – labeled Ctrl A and Ctrl B – which are usually assigned to the Mod Wheel (1) and Foot Pedal (4), respectively, but can be changed to other controllers as appropriate. Again, think about creative and expressive uses of these routings – one favorite technique is to use two spare MIDI-controller faders to fully modulate the positioning of two objects on the string. This can be a responsive way of playing Sculpture, almost like a guitarist subtly moving a pick to various points around the sound hole and combining this with selective damping or muting on the strings.

Figure 7.46
Use the external MIDI controller to manipulate the relative positioning of the objects on the string. This can add a surprising level of expression to your patch.

Figure 7.47
Use Jitter to add pseudo-random modifications and drifts within your sound.

The Jitter, Velocity, and Note on Random modulators can also be used to add variable amounts of chaos, unpredictability, and expression to your sound. Jitter, for example, can create small pseudo-random imperfections in the patch – try adding some Jitter routed to pitch to recreate the slight pitch variations exhibited on some acoustic instruments. Velocity and Note on Random can also be effectively mapped to parameters in the Material pad, so that different velocities, or alternate notes, produce subtle (or extreme) variations in the properties of the instrument. In theory, you could stack all of these modulation sources up, so that Sculpture responds and modulates in a multifaceted and complex way.

Although the modulation paths that we've discussed will start to bring some movement into your patch, probably, the best modulation features of Sculpture are to be found in the envelopes and Morph pad. The envelopes can be used in a conventional way (mapping to Filter Cutoff, for example), or alternatively, to record in modulation data – almost like an automation track within the architecture of the synth itself. To record an envelope into Sculpture, you'll need to activate the small R in the top right-hand corner of the envelope display. This places the envelope into record-ready mode, which, by default, can be initiated into recording by holding down a MIDI note and wiggling the modulation wheel, accordingly. Press R again to stop the recording, and you should see your moves stored into Sculpture's envelope generator.

In essence, the Morph pad stores a series of five snapshots representing positions of the main Sculpture parameters (Object Strength, Inner Loss, Tension Mod, and so on), which can then be morphed using the central XY pad. The great thing here, from a modulation perspective, is that over 20 crucial parameters can be adjusted at the same time (indicated by the small red dots). Now, add the fact that you can record these moves into an envelope to directly form part of the patch, and you have the world's ultimate vector synthesizer!

Figure 7.48
Placing the envelope into record mode will allow you to record your own "movements" into the envelope generator using the modulation wheel.

Arguably, the first step to working with the Morph pad is to define a number of different parameter positions for each of the Morph pad's corners (A, B, C, and D). Clicking on each respective corner, try to adjust the various key parameters, maybe applying differences in the Tension Mods, or the various positions of the objects on the string. The red dots provide reference to the current sound, also indicated by the position of the orange dot on the Morph pad itself. Note that you can also call up the Morph pad's contextual menu (by Ctrl-clicking), so as to copy and paste settings between the different points, or set the randomization destinations (initiated by the Rnd button) to have Sculpture "tweak" the various positions accordingly.

In some situations, you might find it adequate enough to have hands-on control with the Morph pad, and morph the patch as you see fit. Note that the MIDI-controller assignment section to the bottom of the interface details the standard MIDI assignment with Sculpture, allowing you to map the pad's XY axis to two physical MIDI controls on your keyboard. Moving beyond that, though, you might want to record your own vector envelope – unique to the Morph pad – that will then be permanently embedded into the sound. This can especially be effective on pads – with each note taking on its own "wandering" quality as the sound slowly shifts through different phases.

To engage the vector envelope, you'll need to change the Morph pad's mode from Pad to Pad and Env. As with the envelopes in Sculpture, you can record your own moves directly into the synthesizer. In this case, enable the vector envelope Record Trigger (using the small R), play a MIDI note, and wiggle the Morph point accordingly to record the moves in. Pressing R again will stop the recording, with the moves displayed on the pad to illustrate the movement between the various vector points.

Figure 7.49
Using the Morph pad you can define positions for each of Sculptures key parameters, based on your respective XY positioning.

Figure 7.50
Recording your
own vector
envelope will
allow you to
embed Morph
pad movements
directly into your
patch.

7.17   Creative Sampling

Having explored some of the basics of using the EXS24 in Chapter 6, we now turn our attention to some of the more advanced creative aspects of using Logic's integral sampler. The assumption here is that you're using the EXS24 either to map samples (maybe from a sample CD or recorded into Logic), or alternatively, you want to explore some of the deeper mapping and synthesis options within the existing EXS24 instrument library. Either way, you'll find the EXS24 to be a highly developed sound-design tool, and an interesting way of dynamically developing audio files that you've recorded into Logic.

The EXS24 Hierarchy: Zones, Groups, and Editors

As with any sampler, it's important to get to know the various hierarchical aspects of its operation – either in the form of different software pages or the way in which the sampler organizes and manages its sample data into zones, groups, and so on. One of the principal differentiations to make is between the main plug-in window – from which you can select instruments, apply envelopes, filters, and so on – and the EXS24 instrument editor window, which is specifically used for the mapping of samples across the keyboard. For most day-to-day activities, you'll only need to use the main plug-in window; however, if you intend to explore more creative applications of the EXS24 sampler, you'll need to make more extensive use of the EXS24 editor.

Within the editor itself, Logic also makes an important distinction between zones and groups. A zone is the smallest piece of currency in the EXS24 editor, which essentially holds a selected sample mapped across a corresponding range of keys – like C3–D3. A group is a handy way of organizing the various layers of sample data that can be included in an instrument. For example, a group could be used to differentiate between different velocity layers (soft and hard, for example), or two different components of a hybrid patch (like string samples in one group, and piano in another). There're also several interesting functional properties of groups – like the ability to apply different "conditional" triggering options, as well as the option to differentiate alternative filter settings, and so on.

Figure 7.51 The two different sides of the EXS24 sampler – the main plug-in's interface window, and behind it, the EXS24 instrument editor where all the sample data are mapped.

KNOWLEDGEBASE 8

Creating EXS Instruments from Audio Regions

As useful alternative to building EXS24 Sampler Instruments from scratch, consider using the Convert Regions to New Sample Track feature. This feature can work with a number of sounds – allowing you to transform a live drum performance, for example, or even vocals and acoustic guitar, using the power of the EXS24. To convert a region into a sampler track, you'll first need to select the region in question, and then pick Ctrl + click > Convert Regions to New Sampler Track. Logic will then give you the option to create a new sampler track based either on the region's transient markers, or the region as a whole. In most cases, you'll want to pick the transient marker option.

Of course, moving audio regions over to a sampler track is only the start of the creative process, especially given the variety of tricks and techniques possible between the EXS24 and the associated MIDI trigger data. The first option is to explore the various ways in which the source MIDI data can be manipulated and edited – including simple techniques like a different quantize values, or even going so far as to build a completely new "remixed" set of trigger points.

(Continued)

Figure 7.52 Use the Ctrl + click contextual menu option then Convert > Convert Regions to New Sampler Track (or Ctrl + E) as a quick way of creating new EXS Instruments from an existing recording in your project.

On the sampler itself, you can explore a number of different techniques using the EXS24's front panel – including adjustments to the filter, the sample's pitch, and the envelopes. Obviously, the filter can be used to provide a static filtering effect (making the loop darker with a LPF, for example), but things get most interesting when some form of modulation is applied – whether it's note velocity modulating cutoff, ENV 1 shaping each note of the sequence, or one of the LFOs. As Envelope 2 is preassigned to volume, you can create some inserting effects by varying the Decay time, particularly with respect to short "choppy" envelope settings.

If your original audio region contains harmonic or melodic information contained within it, the relative merits of adjusting the EXS24's Tune parameter can be dubious. However, if your source material is percussive, or unpitched, the results can be far more interesting – transforming a clean drum loop, for example, into a crunchy collection of lo-fi beats.

7.18   The EXS24 Instrument Editor

The EXS24's sample editor can be opened using the small Edit tab in the top right-hand corner of the EXS24's main plug-in window – opening up the currently selected instrument in the editor. In many ways, it's a good idea to try opening an existing library instrument first – both as means of getting a grip on how the Instrument Editor presents its mapping and also to see how Logic's sound designers have gone about using the zone and group features to create their instruments.

The EXS24 editor is divided into a series of areas, allowing you to best visualize and navigate the mapping that you've created. First of all is the two main view tabs in the top right-hand corner – Zones and Groups. These allow you to quickly move between viewing the entirety of the sample data and its mapping – zones, in other words – and the "macro-level" properties of a patch organized into groups. For more rudimentary EXS24 patches, it might be that you only need to work in the zones view. However, as your programming gets more complicated, it might become easier to manage by viewing the instrument in both the zone and group view.

Figure 7.53
The Editor interface is organized into a series of areas, allowing you to visualize and edit the mapping of your samples in various ways.

In either view, the screen is divided into four key segments. At the top, you'll find the parameters area, which provides a long list of the samples or groups included within the instrument, alongside their associated parameters (like key range and so on) that define how they function. Moving down from this is the velocity area, which provides a quick snapshot of how samples are placed with respect to velocity – it might be, for example, that a zone or group works over the full MIDI velocity range (0–127) or only across a specific set of velocities (0–40, for example). Below this is the zone/group area that shows the placements with respect to the MIDI keyboard, which is viewable itself in the very bottom of the interface window.

KNOWLEDGEBASE 9
ReCycle Files and the EXS24
Having seen how the Tracks area handles ReCycled files, it's also worth noting the distinct benefits and differences in working with REX files in the EXS24. Put simply, the reason for using the EXS24 over the Arrange page technique is creativity. As we'll see, once the REX files are imported into the

(Continued)

Figure 7.54 Importing REX files into the EXS24 is by far the most creative way of dealing with ReCycle files, facilitating a wealth of addition sound processing and resequencing options.

EXS24, you can start to manipulate the raw data in any number of ways, as well as take advantage of its tempo-tracking abilities by virtue of the loop being cut into a series of slices.

To import the REX file you'll need to be in the EXS24 editor, selecting the ReCycle convert option from the Instrument menu. Arguably the best solution to pick is the Extract MIDI Region and Make New Instrument option, which will not only make a new EXS24 instrument from the audio contained in the REX file, but also place a MIDI region file into your arrangement to trigger the samples in an appropriate way.

(Continued)

Once imported, the world is your oyster! Using the editor itself, you could decide to experiment with some of the zone parameters, with some of the most distinctive treatments coming from the selective reversing of certain slices. On the front panel of the EXS24, you could also decide to experiment with some of the synthesis features, maybe using the filter to make the loop darker, for example, or changing the envelope settings – maybe using a sharper decay setting with no sustain to produce a choppier output. Looking at the region itself, you could decide to resequence certain elements (removing or retriggering certain slices, for example) or even experimenting with different types of quantizing.

7.19 Creating a New Instrument and Importing Samples

Unless you're interested in remapping existing sample data, you'll need to create a new, blank instrument to work with. You can do this through the EXS24's Instrument Editor using the menu option Instrument > New.

Before you start to import sample data, you might want to consider some essential data management issues. For example, if you intend to keep your sample data in a unified place (like an external FireWire drive or a separate partition) you'll want to copy the sample data into an appropriate folder first, before initiating any imports into the EXS24. Of course, it is possible for the EXS24 to reference sample data from any drive – so if you're feeling lazy and have no intention of keeping your data in a unified place, just go ahead and import! Finally, it's also possible to store sample data as part of your project's assets, which might at least facilitate a backup of the sample data, although there's no guarantee that the backup will exist on the drive 3 months down the line!

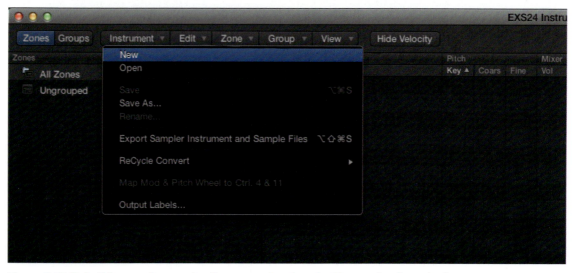

Figure 7.55 To build a sample mapping from scratch, select the New option from the Instrument menu.

The importing process itself can be done in a two principal ways – either through the Zone menu (usually when you intend to load multiple samples in one action) or using a drag-and-drop from the project's Audio Bin and/or the Browser. If you've invested the time in naming your samples properly, or indeed, chosen to import samples from a third-party collection, then the Zone > Load Multiple Samples is probably the quickest and easiest solution. From this option, you'll be able to specify the samples to be loaded, followed by a dialog to provide some indication as to how these should be mapped. The options are relatively self-descriptive, but see the following text for clarity:

Use Auto map for a collection of keyboard samples where the samples span more than one key. In this case, Logic will map the sample intelligently (using the root key included in the sample's name) running each zone into (but not over) the next.

The Drums option again auto maps the samples based on the root key in the sample's name, although this will be the one and the only key the sample is mapped to.

The Contiguous zones is useful where you don't have key information included with the samples, or simply want to ignore any existing key information included with them. Using the Contiguous zones option you'll also need to specify the zone width for each zone (this could be one note, or a whole octave or more) as well as the start note to begin the mapping.

Figure 7.56 Using the mapping option might aid the quick placement of multiple samples – say from a sample CD.

To build the instrument mapping up in a more gradual way, use the Browser window and drag-and-drop the files as you see fit. The exact way the samples are mapped will vary as to the samples selected and how you drag them into the EXS24 instrument editor. Dragging a single sample directly into the zone parameters area, for example, will place the sample across the entire keyboard with its default pitch at C3 (unless specified in the file name). Alternatively, drag a sample onto the MIDI keyboard on the bottom of the editor window, and you'll be able to map it to a specific key, although the range won't extend beyond that single note. Dragging onto an existing zone will also preserve the current zone parameters, but use the newer choice of sample.

Using the browser, you can also select multiple samples and import these directly into your mapping. As with the dialog from the Zone menu, you can also specify how Logic auto maps the samples. Note that by dragging the multiple samples onto a key on the MIDI keyboard, you'll effectively set the starting point, with subsequent zones following on from this point.

Figure 7.57 Besides using the Import option from the Zone menu, you can also drag and drop samples into your EXS24 instrument directly from the browser.

7.20   Changing Zone Properties

As part of the importing process, it's highly likely that Logic will have performed some basic zone parameter adjustments. Let's take a look at the different sections in the zone parameters areas and how you might go about using them.

Key Range

As we've already seen with the process of drag-and-drop sample assignment, each zone is assigned its own unique key range, with both upper and lower settings. If you're creating a drum-based instrument, it's likely that the Lo and Hi setting will be on the same note. However, if you're assembling a multisampled instrument – maybe with two or three samples for every "played" octave – you'll need to adjust the range over a few consecutive notes. You can adjust the Key range through the zone parameters area, or indeed, by dragging the appropriate "mapping blob" in the bottom zones/groups area.

Pitch

Pitch defines the fundamental root position of the note – in other words, the note on the keyboard that plays the sample at its natural sampled pitch. In the above example of a multisample spread over a few notes, you'll probably want to place the key somewhere in the middle of the zone's range, so as to keep the amount of transposition at a minimum. You can also use the Coarse and Fine-tuning parameters to fine-tune the pitch of the sample – maybe the instrument has been recorded slightly out of tune, for example, or you want to explore some more creative aspect of tuning on drum and percussion samples.

Figure 7.58 A drum-based map will place each drum hit on a different note.

Figure 7.59 A keyboard-based multisample might use one sample to cover two or three notes – in this case, the key range should be adjusted accordingly.

Vel. Range

Zones can be set to work over a specific range of velocities, using the Vel. Range parameter. If you've got several samples taken at different velocities, this might be a useful way of achieving a realistic sense of expression in your instrument, although you can always "fake" velocity sensitivity by some careful use of the amplifier and filter responding to velocity.

For more complicated patches – where you might be dealing with several different velocity layers – consider applying the velocity switching through the group feature.

Output

Output allows you to fine-tune some important mix-related issues with respect to your zone. Volume and Pan are largely self-explanatory, but what's of particular interest is the Routing option. By default, each zone is set to the main outputs, but it's also possible to force a zone out to any of the EXS24's 16 individual outputs. To access this, though, you will need to instantiate a Multi Output version of the EXS24 in your instrument channel and activate the additional Aux channels accordingly.

By routing a zone out in this way, you'll be able to gain individual access to audio plug-ins as well a discrete channel fader for the zone(s) in question. This is particularly useful editing tool for drum-based instruments in the EXS24 library that you might want to enhance with additional processing, like compression, equalization, or reverb.

Playback

Playback governs some important aspects about how the sample data is read, mainly in relation to triggering drum samples. For example, the Pitch parameter allows you to enable or disable pitch tracking with a sample, so that even if a drum sample spans several keys, for example, its pitch remains fixed. 1Shot, however, focuses on the duration of the sample, so that in 1Shot mode the full length of the sample is played irrespective of how long the key is held down for. Again, the main application of 1Shot mode is with drums, where the duration of a hit isn't "variable" as such – in other words, a cymbal will always decay over a number of seconds.

"Rvrs" (or reverse) allows you to flip round the playback of the sample data. Given the need to create new audio files before you can reverse region in the Arrange area, this EXS24 reverse function is a quick-and-easy solution in creating reverse effects. Needless to say, the "Rvrs" feature is a great creative tool, sounding great on a range of instruments – from drums hits to pianos and bells.

Sample

Although most samples that you'll import will have their edits already in place, it is possible to further refine the start and end point from within the EXS24 instrument editor. For example, it might be the case that you have a small portion of silence at the front of the sample or that you want to create a more abrupt end to the sample. Although the EXS24 instrument editor displays this numerically, you can Ctrl-click on the sample number to open the sound file in Logic's Sample Editor. If you're running the editor at full-screen size, though, you'll move the EXS24 editor out the way to reveal the Sample Editor.

Loop

The Loop feature is primarily used as means of creating an infinite sustain on sampled instruments like strings, and so on. Obviously, if a sample was taken for the full-note duration, this could make an exhaustive drain on RAM allocation (although some third-party libraries do this), so samples are taken of about 5 s or so in length, with the last 3 s looped over until the notes end.

In the case of the EXS24, you can define a number of parameters in relation to loop, with the principal aim of creating a smooth, glitch-free loop that shouldn't be particularly audible to the ear. The beginning point is always defining the Start and End point of the loop – ideally picking a section of the sample that doesn't waver or drift too much, but instead, has a clearly defined sustain. Although you can define a loop point numerically, the easiest way to do this is to open up the Sample Editor (by Ctrl-clicking on either the start or end parameter number) and perform this visually. Make sure that you've got the Search Zero Crossings enabled (under the Edit menu), so that any start and end points snap to a zero crossing point.

Figure 7.60 Although it is possible to set the loop numerically, opening the loop points in the sample editor (by Ctrl-clicking them) will allow you to better place the start and end points to match what is happening in the waveform.

Even if you do find a relatively successful location to position the start and end points, it might still be important to use some of the EXS24's additional loop parameters to fine-tune the looping to produce an even smoother looping effect. Xfade, for example, can be useful if you're still hearing small clicks or jumps around the loop point, with a larger number creating a longer crossfade. E. Pwr further refines the crossfade, so that no energy is lost in the crossfade processing, while tune addresses any tuning issues that might have become evident in the looping process.

7.21   Working with the EXS24's Groups

As soon as you start building up instrument files with more than ten or so zones, it becomes essential to start to manage the samples in some meaningful way. Groups offer a variety of benefits to the EXS24 sample list – from organizing better ways of displaying related zones, to advanced triggering options. Using the instrument editor you can create groups in much the same way as you might create playlist in iTunes – simply select the zone or zones you want to group and then drag these across to the zone column on the left.

In a rudimentary way, the zone column offers an excellent way of navigating your instrument – either viewing the complete set of zones in an instrument or choosing to view the instrument group-by-group

Figure 7.61 Select the zones you want to place in the group then drag them over into the zone column to create a corresponding group.

by clicking on the appropriate group folders in the zone column. Note that you can also name the groups by double-clicking on the group name. This is also an excellent way of organizing and creating hybrid EXS instrument – selecting different layers of zones from one instrument and copying and posting them into another.

Advanced Mapping Options

Moving over to the groups view in the EXS24 instrument editor, you should also see a number of parameter options unique to groups you have created. What groups offer is a form of macro-level control over an instrument – so, rather than having to individually set velocity switching over hundreds of zones, we can place them into four respective groups and then simply alter the velocity setting of those individual groups. Not only are grouped velocity adjustments far quicker in the short term, but also for any adjustments later on, the group feature is invaluable.

As you'd expect, the Key Range, Vel. Range, and Output parameters function in much the same way as they did in the zone parameters, although of course, settings here override above what you've established in the zone (for example, if the group's key range doesn't extend as far as where you placed a zone, you won't hear the sample).

The Trigger option defines whether a group is triggered by a Note On or a Note Off command (known as Key Down and Key Release, respectively) – an effect more broadly referred to as Release triggering, which uses short samples of events that occur after the main note to create a more realistic and life-like "musical event." Release samples could be in the form of reverb, for example, or the small decaying "resonance" when a piano note is released.

Figure 7.62 Groups contain a unique series of mapping options – including filter and envelope offsets and various conditional triggering options – ideal for the expert EXS24 user.

Following on from the trigger options – although positioned over the far right-hand side of the group parameter – are the conditional Select Group By options. Although you can create a number of different conditional triggering options here, probably the most well-known is the option to create a so-called key-switched instrument. With key switching, a number of notes down the bottom of the keyboard will be used to dynamically toggle between different instrument articulations (like staccato and legato), all on the fly. Each articulation, as such, is organized into its own group, with the group "selected by" a specific note, like C1.

The Select Group By actually works with a number of different MIDI events (Note, Controller, Pitch Bend, and Channel). This makes conditional triggering an incredibly flexible feature – allowing you to use a modulation wheel, for example, to switch between layers.

The Envelope and Filter Offsets allow you to differentiate between the various groups and the initial settings on the EXS24's front panel (which we'll look at in more detail later). For example, although you might set a master filter cutoff position at 50%, for example, you could use the Filter Offsets parameter so that one group opens the filter in a more open position ±50), while the other closes in down (−25). Although this isn't quite the provision of a completely independent multimode filter per group, it does allow different zones to use different filter settings concurrently – in other words, while one group sounds filtered, the other is less so.

Following this concept further are the envelope offsets, which again work from the basic premise of the parameters established on the main front panel, but then adapted, negatively or positively, to match the specific needs of the group. You could, for example, have a nylon-string guitar in one group using the fast attack and release settings, while in another group, some string samples use softer attack and release settings to produce a lush background pad.

7.22   Editing EXS24 Instruments Using the Front Panel

With the samples mapped, now comes the task of adjusting and refining the synthesis engine of the EXS24's sample playback. The settings on the front panel extend what you've already achieved in the mapping of the samples. For a simple instrument, for example, all you might need to adjust is a few key parameters like release, or its velocity scaling. Alternatively, you could use the front panel of the EXS24 to completely transform the samples way beyond their original sound – maybe using some extreme filter and drive settings, combined with a series of creative modulation-matrix settings.

Like the ES2, the front panel is divided into a number of key sections, including the filter block, the modulation matrix, and the LFO and envelope modulators. To understand how these parameters work, let's look at a couple of different editing scenarios and how these might work.

Basic Instrument Refinement: Velocity Scaling, Polyphony, and Envelopes

Not all instruments require a great deal of modification away from the default settings that appear on the EXS24's front panel, but there are a couple of areas worth looking at to best optimize the instrument in the final mix. First of these has to be the volume controls over the right-hand side of the plug-in interface. Alongside the overall volume level (which might be worth reducing if you're using a lot of sample voices simultaneously), you'll find the sampler's scaling of level through velocity. Raising the bottom part of the level through velocity slider can sometimes be a useful alternative to compression, especially if you've got a multisampled instrument with "quieter" samples in its low velocities.

Adjusting Envelope 2 – which, by default, is assigned to the Amplifier duties – is a useful way of quickly grabbing hold of, and shaping, an instrument's ADSR characteristics. For example, if some of the samples in your instrument have reverb embedded into them, you might find it beneficial to increase the release time so that you hear the full decay of the room ambience. Alternatively, using a slower attack can be a valid sound design treatment to soften or disguise the initial attack transient of a sample.

Toward the top left-hand corner of the interface are a number of useful macro controls, some of which can have a big impact over the performance of the sampler. For example, the voices parameter reserves a given amount of polyphony for the instrument in question. Sixteen notes, for example, might be more than enough for a simple drum-based instrument, but should you use something like a multisampled piano with release samples, you might need to raise this up to the full 64 voices. However, with an increased voice count comes an increased CPU drain, so it's often the case that you actively try to balance out the realistic demands of the instrument, against the overall needs and drain of your CPU performance. A used indicator at least provides some rough indication as to the total used polyphony, but you should also be able to hear a "maxed-out" polyphony by longer notes appearing "snatched off."

Adding in the Filter

As we saw with the ES2, the filter has a surprisingly powerful way of shaping the timbre of our sound source. The filter included in the EXS24 is certainly no exception to this rule: with up to six different modes of operation (including high-pass, BP, and four different strengths of low-pass), some fat and juicy resonance, and even a Drive parameter.

Figure 7.63 The main controls in relation to volume are on the right-hand side of the EXS24's interface. Try experimenting with the sampler's response to velocity, as well as how Envelope 2 changes the amplitude over time.

To activate the filter, you'll need to press the On switch, otherwise it defaults to its off position so as to save a few CPU cycles. Switching the modes can be done using the six tabs at the bottom of the filter, with the additional Fat button to preserve some of the low-end with high-resonance settings on the LPFs. Remember that the precise position of the filter cutoff and resonance can also be varied between different groups, using this front panel setting as the initial starting point.

Modulation Heaven!

Many of the more advanced synthesis options come through the application of the modulation matrix, which allow you to map a variety of sources (including the LFOs and envelopes at the bottom of the interface) to various destinations within the EXS24's synthesis and sampling engine. As with the ES2, the EXS24 uses a familiar matrix display, with a Destination parameter (like Filter Cutoff), and amount of modulation and the Modulation Source including any of the EXS24 modulators (three LFOs and two envelope generators) or any one of the 127 standard MIDI controllers.

Figure 7.64
Once you've engaged
the filter, you can use
its variety of modes
(low-pass, band-pass,
and high-pass) to
shape the timbre of
the EXS24's output.

Figure 7.65 As with the ES2, a large Modulation Matrix section allows you to route a number of different modulations sources to targets in the EXS24's programming architecture.

Saving Instrument Settings

Confusingly, the EXS24 makes some level of distinction between the instrument settings stored with and without the front panel controls. To permanently store the front panel setting alongside your instrument mapping, you'll need to go to the Options menu on the EXS24 and select the option "Save Settings to Instrument." Now, when you return to that instrument, the mapping will be loaded alongside any front panel settings that you've created. Alternatively, use the Recall Settings from Instrument or Recall Default EXS24 Settings if you've made changes to the front panel settings that you're not happy with.

Figure 7.66 If the settings on the EXS24's front panel have become intrinsic to the instrument in question, consider using the Save Settings to Instrument feature under the Options menu.

7.23   Ultrabeat

Ultrabeat is Logic's take on a contemporary drum machine, complete with a range of synthesis and sequencing options, as well as a handy multichannel version to allow the flexible access to Logic's signal-processing features. An Ultrabeat Kit is made up of 25 different drum voices – listed down the left-hand side of the interface complete with level controls (as the blue bars), mute, solo, pan, and an output selector. On selecting a drum voice, you can then edit its associated synthesis parameters in the large central display, or indeed, program its unique step sequence at the bottom of the plug-in.

In keeping with the flexibility of the ES2 and Sculpture, the synthesis options in Ultrabeat are impressive to say the least! Each drum voice has access to up to three different oscillators, which are then passed through a multimode filter, a ring modulator, and finally some EQ. Arguably, the most useful oscillator, though, is Oscillator 2, which features a sample playback mode as well as the phase Osc and model modes. On the whole, you'll find that samples offer the most quick and effective solution for assembling

Figure 7.67 Ultrabeat features up to 25 different drum voices in each kit, each with their own set of synthesis options.

a kit (if so, use the "blank" Drag & Drop Samples Kit from Ultrabeat's plug-in settings), although it's also good to augment these with some of the pure synthesis elements, like the noise generator. Once the sample mode is active, you can also drag sample files directly from the browser onto Oscillator 2.

Clicking on the Filter tab activates the filter, offering a great way of coloring the samples, like a BP filter applied across a clap sample to make it sound narrower. To modulate the filter, you can also use any one of the four envelopes and an LFO, with the small blue legend beneath the cutoff and resonance controls, allowing you to define an appropriate modulation source. If you're after grit, though, try activating either the crush or distort modes toward the bottom of the filter dial – both effects that can sound great on heavier kick or snare samples that you want to push to the front of the mix.

Figure 7.68 Use the Browser to drag and drop samples directly into Ultrabeat's second oscillator.

When it comes to assembling the sequence, you can just step through each voice and create the finished pattern piece-by-piece. A much more effective solution, though, is to switch Ultrabeat into its Full View mode using the small tab in the bottom left-hand corner of the interface. In full mode, the synthesis controls are removed to reveal a full-sized drum grid for all of Ultrabeat's 25 drum voices. This makes it far easier to gauge the interaction between the various parts, rather than having to remember their various hit points. Note that to hear the sequencer in action, you'll need to activate its Power tab, with the transport controlled either from Ultrabeat itself (in which case it plays independent of the rest of the project) or from Logic's main transport, so as to audition the pattern in the context of the song.

In total, you can sequence up to 24 different patterns with Ultrabeat, with the potential to switch between them using a MIDI keyboard. A much better solution, though, and one that makes more sense in the Arrange area, is to export the pattern out as a region. To do this, click-and-drag on the small icon next to the pattern selector, placing the resultant region on Ultrabeat track in the Arrange area. Exporting a number of patterns in this way, you can then revert back to Arrange area to structure your song accordingly, even using other MIDI editors (like Piano Roll or Hyper Editor) to further manipulate the data.

Figure 7.69 Use Ultrabeat's Full View to sequence the complete pattern, which can then be dragged directly into Logic's Arrange area.

In this chapter

Mixing in Logic

8.1 Introduction

Up until a few years ago, there was only one accepted way of creating a professional sounding mix: hiring a commercial studio facility and mixing through a traditional mixing console, like an SSL or a Neve. However, times have changed significantly since then, and mixing "in the box" – in other words, mixing completely within the domain of your audio sequencer – has become a viable, and some would say more flexible, alternative to the traditional console route. In truth, applications such as Logic Pro offer a tremendous amount of creative and technical freedom to today's mix engineer: complete and instantaneous recall, compression and equalization on every channel, studio-grade reverb, full automation, and much more besides.

In this chapter, we're going to take a closer look at the process of mixing in Logic Pro, looking at both the technology it has to offer to create an effective mix and how we can knit these elements together as part of the mixing workflow. This is also an opportunity to get to know many of Logic Pro's essential signal processing plug-ins in a more informed way, as well as taking a more detailed look at how the Mixer area works – aspects that can actually improve the entirety of your workflow in Logic Pro, as much as the mix. Looking more closely at the process of mixing itself, we'll explore how to use compression and equalization to create separation, as well as a bus processing (on aux channels), and of course, spatial treatments like reverb and delay that help define the depth and dimension of your mix.

Beyond the essential signal processing tools of mixing, we'll also take a look at features like automation, grouping, and folders, all of which help make your mix easier to manage, especially with the large track counts that most projects seem to encompass.

8.2 Channel Strips: Understanding Your Virtual Console

Before diving headlong into your first mix on Logic Pro, it's worth taking some time to distinguish among the various channel strips that the Mixer is comprised of and how these various elements interact with each other. Generally, a good mix in Logic Pro makes use of all the features of the Mixer – channel inserts, bus sends to aux channel strips, output masters, and so on – so it's worth familiarizing yourself with their features in much the same way as an engineer would "get to know" the sections of a physical console.

Audio Channel Strip

An audio channel, as we've already seen in the previous chapters, governs the basic path in and out of Logic Pro for a recorded signal. In a mix, audio regions in the corresponding track will be sent down the channel strip, through the various insert processors (compression and EQ, for example) and bus sends (for reverb and other effects) to a designated output.

Figure 8.1
An audio channel strip.

Figure 8.2
An instrument channel strip.

Instrument Channel Strip

The instrument channel strip duplicates the same features as an audio channel strip; only this time you get to work with the signals generated by the virtual instruments in your session. In addition to this simple application, an instrument channel strip can also be configured for multichannel operation so that multiple outputs from the same instrument (like the EXS24 or Ultrabeat) are sent to a number of additional auxiliary output channels. Besides having individual level control, this also allows you to apply different effects onto each output so that a snare, for example, might have different equalization, compression, and reverb to that of a kick drum.

PLUG-IN BOXOUT 1
Distortion Effects

Logic includes an impressive set of effects dedicated to low-fi transformations and crunching-up sounds – from the digital extremes of Bitcrusher to the warmer tones possible with Distortion II to the detailed emulation of many favored tones emulated within Amp Designer. Besides being an obvious addition to guitar parts, distortion can also be a great source of color and interest on any number of sounds on a mix. Try using distortion to add some grit and body to a drum loop, for example, or a touch of drive on an aggressive vocal.

The first choice of amp-like distortion would have to be Amp Designer, which provides immense flexibility with respects to the choice of emulated head, speaker cabinet, microphone type, and its position. Additionally, there are a considerable wealth of presets to get you stared. Add to this the power offered by the new Pedalboard plug-in offering the guitarist endless scope.

Distortion, overdrive, and clip distortion come from earlier incarnations of the application and tend to be less suitable for applications on electric guitar. They are, however, an excellent addition to drum loops or acidic synth lines generated from the ES2. Both distortion and overdrive have particularly simple controls with just a drive and tone parameter, alongside a corresponding output reduction slider to avoid ripping your monitors to shreds!

The prize for the most bizarre distortion effects goes to the Phase distortion and Bitcrusher plug-ins. Phase distortion sounds great on drum loops needing to be mashed up beyond all recognition. Based on a modulated delay line, it can produce tones that range from a warm fuzz to the sound of a FM broadcast gone considerably wrong! Bitcrusher, on the other hand, is a great source of digital grit – especially as you reduce down the bit slider (to around 8 bits) and start applying some heavy downsampling (103 and beyond).

(Continued)

Figure 8.3 Distortion can bring plenty of color and grit to your mix. Try experimenting with Logic's wide range of distortion effects to achieve anything from a subtle valve-like warmth through to early digital distortion.

Aux Channel Strip

The humble aux channel strip is one of the most useful parts of Logic Pro's Mixer and the source of a number of different techniques in mixing. First, it can be used as means of applying send effects like reverb or delay – anything where you want to actively control the balance of wet and dry sound using

Figure 8.4 An aux channel strip.

a separate fader. In this example, the aux channel strip has the effect strapped across its insert path with instrument channel strips or audio channel strips sending signals via designated bus sends. This allows any number of channels to access the same reverb, for example (an essential way of preserving DSP resources), as well as being able to control the return of the effect in the same way as any other audio signal in the mix – in other words, it could be equalized, compressed, and controlled by movements of the fader.

Another less immediate, but altogether just as useful, technique is to use an aux channel strip to combine a number of different channels (including audio and instrument channels) on a single fader. For example, many engineers using a traditional console will create a submix of the drums to a selected bus master fader (in this case, one of Logic's aux channel strips), allowing them to quickly control the level of the drums relative to other instruments in the mix. Additionally, they could also make use of inserts on the bus faders to apply compression, EQ, and so on, to the entirety of the drums, rather than individual channels within it – something which can just as easily be done of Logic's aux channel strips.

The final important role of the aux channel strip is to operate as a means of inputting external signals into the mix. If your audio interface supports enough inputs, there's no reason why you can't patch-in external compressors, equalizers, synthesizers, and so on, directly into your mix. Although you might need to account for a small amount of latency (as discussed in Chapter 3), using dedicated recording hardware can often supply much more character than conventional plug-ins. Of course, given the provision of inserts on these aux inputs, there's no reason why you can't also use Logic plug-ins on top of whatever device you're inputting into that particular aux input.

Stereo Output Channel Strip

The stereo output channel strip represents each of the physical outputs in your Logic system. If you're using a simple two-in, two-out USB audio interface, you'll only see one of these faders. However, if you're working with a multiple output FireWire or PCIe soundcard, you can increase the number of output

Figure 8.5
An output
channel strip.

faders corresponding to each physical output. On the whole, most mixes simply use one designated output (1–2) as the "destination" of the two-track mix. If you intend to render the mix directly from the Logic session, you'll also need to make use of Bounce "Bnce" button, found on the output channel strip or File > Bounce (for more information on bouncing to disk, see Chapter 9).

PLUG-IN FOCUS
To Gate or Not to Gate ... That Is the Question

Music trends change, so it's not a surprise that the process of recording also adapts concurrently. For example, just a few years ago, it was commonplace to gate the components of a drum mix (kick, snare, toms, and so on) to within an inch of their existence! Nowadays, it seems that listeners, engineers, and musicians all seem to prefer a looser sounding kit, resulting in much less use of gating.

Now and again, though, you may well feel the need to improve your aural hygiene – maybe a vocal microphone has some obtrusive background noise made worse through the application of compression or that one and only guitar take has a little too much noise from the amp. If you're pushed for time, there's no doubt that the noise gate plug-in remains a quick and effective way to attenuate problematic noise in between notes – although remember to gate before the compressor and not after it! The trick with setting the noise gate is to start with hard settings (quick Attack, Hold and Release, with 100 dB of reduction) to find the right threshold. Although the gating sounds harsh,

Figure 8.6 Two choices for improved aural hygiene – the traditional noise gate or the current favourite, Strip Silence.

(Continued)

you'll be better able to find the "sweet spot" just above the amplitude of the noise. With this set, back off some of the settings, especially Release and Reduction to soften the effect – surprisingly, even a small amount of reduction (6–10 dB) can have a big effect on the overall cleanliness.

If you've got a little more time, most users now tend to approach the gating issue through a few crafty edits. Try using Strip Silence (Ctrl + X) as a means of Logic preparing the majority of edits for you. As with a conventional noise gate, you'll need to establish the right threshold to get the best results, although arguably this is easier to "see" on Strip Silence than it is to "hear" on the noise gate. However, the big advantage with the Strip Silence approach is that you can modify the edits later on, making sure none of the important attack transients are missed out.

MIDI Channel Strips

Arguably, MIDI channel strips should be kept distinct and separate from the audio mixing, but given their inclusion as part of the Mixer, it's worth clarifying their exact role. Unlike all the other Mixer channels in Logic (which control the flow and qualities of audio within the application), the MIDI channel strips have no direct control of the "internal" properties of your mix, but instead they work as controllers for external MIDI hardware connected to your MIDI interface. However, in the case of the synths and sampler being returned via aux channel strip, these MIDI channel strips could have a direct effect on your audio mix – maybe balancing 16 MIDI channels being returned to one stereo input on your audio interface.

Figure 8.7
A MIDI channel strip.

Master Channel Strip

The master channel strip provides a quick-and-easy access point to the level of signal going to all main outputs and is directly linked to the "volume" control as part of the features on Logic Pro's transport bar. In situations without control room monitor levels, for example, this could be used as a means of adjusting the monitoring level, although it should be noted that any digital bounces made in Logic will be subject to the master channel strip's gain adjustments.

Note that you cannot insert plug-ins across the master fader in stereo mixing (instead use the Stereo Out strip), with the master channel simply being used as a level attenuator. In surround mixing, however, its role – and the role of the outputs – slightly changes, which we'll cover in more detail in Chapter 10.

8.3   Organizing Your Mixer: What You Do and Don't See

Ask any good mix engineers about the fundamental tool that helps them negotiate an effective mix workflow, and they'll often reply "a well-organized mixer." On a traditional console, for example, engineers might decide to repatch the tape returns so that the order and arrangement of the various channels best reflects how they might want to carry out the mix – maybe with the drums in the first 8–10 channels, for example, while bass, guitars, and keyboards following on subsequent channels. So before you start diving into equalization settings and designing your soundstage, take some time to organize both the arrangement and track list to help navigate your way around the mix successfully. Logic Pro also provides a number of ways to organize and structure the appearance of your mix, for example, separating off sections, hiding unwanted channels, and viewing the complete signal path.

Mixer Views

The first concept to grasp with respect to using the Mixer interface effectively is the link between the Tracks area and the Mixer. By default, the Mixer is set to its Tracks view out of the three possible view modes at the top of the Mixer window – Single, Tracks, and All. In Track mode, the order and amount of channels displayed directly corresponds to the Tracks area track list. For example, you might have reorganized the vocals to place them on the first track lane – a decision that also impacts on the arrangement of the Mixer. So, one simple way of organizing your Mixer is to organize your track list!

Following this concept, you'll notice that the track Hide feature that we first saw in the editing chapter also carries through into the Mixer. Tracks hidden in the Tracks area, therefore, will be temporarily removed from the mixer view.

An alternative to the Tracks view is the All mode, which displays the entirety of audio channels in your Logic Pro project arranged sequentially by their assignment (Inst 1, Inst 2, and so on), rather than their order in the Tracks area's track list. This is a useful way of seeing the entirety of your mix, including objects (like the metronome and Prelisten channel) that you wouldn't usually have access to. However, without the ability to organize this view, it is easy to become disorientated, especially in bigger mixing sessions.

Figure 8.8 and 8.9 In Arrange view mode, the Mixer area directly corresponds to the order of the channels in the track list. The Hide Track option can also be a useful way of removing further channels.

KNOWLEDGEBASE 1

Clipping Faders: Good or Bad?

On the whole, conventional engineering wisdom tells us to avoid situations where a signal "clips" the meters – in other words, whenever we see one of Logic's channels faders or master outputs go into the red. However, in reality, there are some situations in Logic where this is acceptable and others where it demands your immediate attention!

In Logic, a fader registers a clip whenever it is presented with too much signal – this might be because you've brought up the output gain of the compressor too much, or that you're feeding too much signal to the main output channels. Looking first at channel faders and instrument tracks, the clipping we see here isn't particularly problematic, with no immediate chance of you hearing distortion at first. This is all because Logic's Mixer incorporates a degree of safety margin – or headroom – built in, using an enhanced bit-depth resolution to tolerate peaks in the region of 11–16 dB.

Figure 8.10 Although clips on channel faders aren't too problematic, it is advisable to avoid clipping the main output.

Moving onto the main output channel strip, however, things are not quite so flexible. Any final digital medium has a fixed amount of level it can tolerate – in other words, a CD or WAV file cannot exceed 0 dBFS (0 dB on the output fader) without the top of the waveform being clipped or distorted. In this respect, it is important to ensure that the main output fader doesn't ever go into the red when you print off your final mix (although the odd peak during the mixing process isn't too disastrous). Given the amount of signals potentially feeding this channel, though, this is quite easy to encounter.

So, if your output channel does clip, what should you do to resolve this? The first and easiest solution might be to turn down the faders contributing to the mix, which could be done by placing them all into a group, and then turning the group down accordingly. If it's in the region of a few decibels, you could also choose to turn the master channel strip fader down; although if you're using it in the realms of 2–6 to 2–10 dB, then it suggests that there is a fundamental level mismatch that needs to be addressed. Another option is also to put an instance of the gainer plug-in (on the master output) and also use this to reduce to final mix level ahead of it being bounced.

Viewing by Signal Path

One of the most interesting view modes is the so-called Single mode, which shows the complete signal path of a selected channel. The signal path illustrates the full journey a channel makes through Logic Pro's Mixer, including all the associated bus send effects, any submixing via aux channel faders, and of course, the final output channel strip. Using Single mode is a great way of honing in on the specific signal-processing characteristic of one sound in the mix, especially where it's using multiple bus sends (for reverb, and so on) that could be positioned well along the Mixer in the standard Tracks view.

Dropping Out Sections of Your Mixer

Using the various tabs along the top right-hand corner of the Mixer area, you can switch out various corresponding sections of your Mixer. For example, you might choose to remove MIDI faders given that they don't directly input on the signal processing aspects of the mix. Alternatively, by removing the audio and instrument (Inst) from the view, you can gain quick access to the aux masters and main output without having to scroll across the full length of the Mixer.

Figure 8.11 In Single mode, you can see the complete signal path for the currently selected track.

8.4 Summing Stacks and the Mixer

As we saw in Chapter 5, Track Stacks are useful in keeping the Tracks area and mixer relatively clutter free. For example, you might decide to pack away your drum parts into an individual Track Stack, keeping the individual mix channels (Kick, Snare, Hi-Hat, and so on) away from the main view of the track.

Figure 8.12
Use the folder function to "pack away" parts of the mix. This can be useful in projects with large amounts of channels or tracks.

PLUG-IN BOXOUT 2

Helpers ... Handy Little Plug-Ins

Sometimes the small, innocuous little plug-ins can be as vital to creating an effective mix as the big processor-hungry reverbs. Take the Gain plug-in, for example, its main application is simply to raise the gain of an input – surely not that useful then? Say you've recorded in a series of complicated

Figure 8.13 Many of Logic's smaller plug-ins can aid the effective workflow of a mix – from the Gain plug-in to the range of metering options.

(Continued)

automation moves but simply want to raise the entirety of the track by 2 dB. Usually this would necessitate editing or rerecording the moves. Alternatively, you could use the Gain plug-in (patched somewhere in the insert path) and simply select a 2 dB boost! Some of its other applications can also be great on stereo signals – switching the left- and right-hand sides, for example, or mono'ing a stereo mix (try leaving an instance across the stereo bus) to check its mono compatibility. Also, if you need to phase invert any microphones (a snare bottom, for example), simply activate the Phase Invert option.

The Multimeter plug-in is a great addition to any stereo bus, providing up-to-date information on the track's spectral properties (using a 1/3 Octave Spectrum Analyzer), alongside phase characteristics courtesy of a distinctive Goniometer. Use the spectrum analyzer to get a better grasp on the spread and range of frequencies in your mix. If you're using monitors with a limited bass response, the lower spectrum information (100 Hz and below) can be crucial to keeping an eye on any excessive activity. The Goniometer provides some indication of the stereophonic properties of your mix – a strong mono image is indicated by a clear line up in the middle of the display, while excursions into stereo are indicated by movement to the left and right.

The Correlation Meter – available as a separate plug-in or, on either of the Goniometer or Analyzer's screens – indicates phase problems with the mix. If the phase is good, you should see consistent movement in one side of the line (usually in the 11 region). However, if the phase is bad (indicating problematic mono compatibility), the Correlation Meter moves between both −21 and 11 in a continuous fashion.

When it comes to creating Tracks Stack, you have the choice of either creating a Folder Stack (as we saw back in 5.30 Using Track Stacks, Folders, and Hiding Tracks) or a Summing Stack. For mixing purposes, a Summing Stack is a useful alternative to the Folder Stack as the function creates both merged track lanes alongside a new auxiliary buss so that the stack can be controlled from a single fader. The advantage of a Summing Stack, therefore, is that we can add additional processing – like compression and equalization – that particular group of instruments, something that we'll explore in greater depth in section 8.7, Combined Processing Using Aux Channels.

8.5   Beginning a Mix

With the mechanics of understanding how Logic Pro's Mixer works under your belt, let's start to have a look at the creative and technical processes of putting the mix together. However, before you start instantiating various plug-ins, take time to establish a plan for what you want to achieve. Think about how you intend to distribute instruments across the soundstage – in respect to both left-to-right placement (pan) and front to rear (with the use of reverb and maybe EQ). Establish what you feel should be the "lead" instrumentation – in other words, the three or four instruments that really hold the track together – alongside the other elements that should only provide a "supportive" role. Finally, identify instruments that are possibly fighting against one another, sounds that may blur the definition and energy of the track, or parts that may get lost among other instrumentation. Ultimately, if you're clear about some of these issues from the start, you'll have a much better chance of producing a coherent end result.

Basic Part Leveling

An essential part of the mix revolves around good levelling practice – put simply, balancing parts without the addition of any additional processing. Although the precise technique varies from engineer to engineer, the intention is to get a good "working balance" of sounds, some with more dominant roles than others, but all without overloading the main outputs from Logic. For example, it's easy to start putting a mix together, pushing channels into the 13–16 dB region only to discover that the main output quickly starts to distort. Instead, try to keep an eye on the main output level (usually on the output 1–2 channel, which should sit side-by-side the instrument channel/audio channel in the inspector) and bring in two or three of the key instruments leaving 6–12 dB or so of headroom. Unless an instrument has been recorded particularly "hot," this will probably be in the region of −24 to 0 dB.

Alongside the basic of part leveling, you might also want to consider panning the various channels to create a realistic soundstage. Think about the instrumentation arranged across the stage, creating a natural, well-distributed configuration of the instrumentation, without relying too much on the extremes of left and right (−64 and +63, respectively) and leaving a clearly defined space in the center of the mix for important lead instrumentation like vocals (in other words, only a few channels should be left in the 0 position). Remember that as you apply effects like reverb and delay (which we'll see more of later), you can still pan these signals just like any other channel in the mix.

Figure 8.14 Try using equalization to create better separation between instruments. In this example, a 3 dB boost at 3.5 kHz in one guitar has been balanced out by a cut in 3 dB at 3.5 kHz on another.

Applying Equalization and Compression

Two of the most important signal processors in Logic – and the cornerstone of any professional mix – have to be the equalization and compression plug-ins. One of the main applications of compression and equalization is to provide better separation between sounds – fixing different elements into specific parts of the frequency range of a track, for example, or locking a sound better into the mix's dynamic properties. One common technique with EQ is to reduce clashing frequencies between sounds, allowing each sound to sit in its own sonic space. For example, a 200-Hz reduction on a low guitar part would facilitate a little more space for a bass, whereas the intelligibility of the vocal could be improved by a cut at 1–3 kHz on any competing instruments. Compression, on the other hand, helps iron out any dynamic inconsistencies – for example, rather than a vocal dipping in and out of the mix, it can remain consistently loud.

Compression and EQ both can be applied via the insert path of a selected audio channel strips, instrument channel strips, or aux channel strips, although the EQ can also be quickly activated by double-clicking in the small EQ box at the top of the channel strip. Double-clicking the selected plug-in will then open a floating dialog, allowing you to adjust the various parameters contained within the plug-in. Of course, bypassing effects is essential to establishing exactly what you have or haven't achieved. This can be done

PLUG-IN BOXOUT 3

Compressor

Along with EQ, compression forms one of the fundamental tools of a mix-down and essentially works as a form of automated amplitude control – the compressor reacts to loud peaks in the input and then attenuates the level of the signal accordingly. The two key parameters in this process are threshold – rated in decibels and governing the point at which compression begins – and ratio, which sets the hardness of compression applied. For example, with a low threshold and ratio (−220 dB and 2:1, respectively), a slight compression would be applied, even on relatively quiet parts of the signal. Alternatively, a higher threshold and ratio (−25 dB and 10:1, respectively) would produce a stronger compression, but this would only be applied on the rare occasion the signal level exceeded −25 dB.

In addition to the key compressor settings (threshold and ratio), Logic's compressor plug-in contains a number of other parameters to further refine the processor. Attack and Release, for example, are important tools for defining how the compressor moves in and out of gain reduction, as the signal exceeds the threshold. With a quick Attack and Release setting, the compressor is set in a "fast-acting" mode – the response to loud transients (like loud snare hits) is quick and efficient, and the compressor is quick to return to its normal state once the input reduces below the threshold. As great as this sounds, however, these fast movements in and out of gain reduction can have a negative effect on sound being processed, as the compressor creates a distracting "pumping" sound. Setting a slower Attack and Release might allow the odd loud peak to slip through the net, but you'll end up with a more musical, empathetic response from your compressor.

Peak and root mean square (RMS) modes govern how the compressor "listens" to its input. Essentially, the Peak mode responds exactly to the true level of the input, while RMS responds to an averaged

(Continued)

Figure 8.15
Logic's compressor is a versatile gain reduction tool, capable of controlling the precise amplitude and dynamic range of a signal in the mix.

level, closer to how the ear perceives loudness. With a Peak detection setting, therefore, you'll find the compressor reacting more than on the corresponding RMS setting, although its response might be considered slightly less musical. The Knee of the compressor is a useful means of creating a more graduated transition on harder ratio settings – the ratio is slightly softer (2:1, for example) ahead of the threshold, only reaching its full strength (6:1) a few decibels after the threshold.

As the overall effect of compression is to reduce the dynamic range of your input (in other words, making the loud bits quieter), you'll probably find your corresponding output quieter than without compression. Increasing the output Gain (at the right of the interface) restores levels lost through compression. You can also use the AutoGain option of Logic to apply this automatically, although in some cases this can produce distortion.

using the dedicated bypass control in the top left-hand corner of the plug-in's interface or directly from the Mixer itself by option-clicking on the appropriate plug-in slot.

Adding in Further Plug-Ins

Further plug-ins can also be inserted as part of the path of plug-ins you create on the channel strip. Remember, though, that the order in which you insert the plug-ins can have a big effect on the overall treatment produced. Even something as simple as EQ and compression, for example, can have a subtly different output based on which plug-in is placed first in the chain. More extreme effects, say, for example, distortion followed by EQ or EQ followed by distortion, can have completely different result given their relative position. If you need to change this order, you can use the Hand tool on the Mixer (or simply press Cmd + click the plug in) to reorder the inserts in any way you see fit.

Figure 8.16 The order in which plug-ins are inserted can have a big effect on the end output. Try using the Hand tool to experiment with the order of your plug-ins the channel strip.

One common mistake at this point is to apply too many plug-ins, possibly in a desperate attempt to improve apparent deficiencies in the source recording; "at least if it's covered in distortion and delay, nobody will notice that it's out of tune!" Try to have a clear strategy in your application of effects – not all tracks necessitate the use of plug-ins, and where a more complicated series of plug-ins are used, try to make this a unique and identifiable "special" feature, rather than the norm.

PLUG-IN BOXOUT 4
Channel EQ

The Channel EQ is easily one of the most important plug-ins in any mix, controlling the precise spectral qualities of signals passing through it. Although the Channel EQ takes a low drain on available CPU resources, it is both a powerful and a flexible tool, split into eight frequency bands with a cut/boost and frequency parameter for each band. As a good starting point to understanding your input source, try activating the Analyzer feature of the EQ. The Analyzer provides a real-time Fast Fourier Analysis of your signal, indicating the distribution of energy across the sound spectrum – a bass, for example, should produce large "humps" formed by its fundamental at 100 Hz, alongside additional harmonics further up the harmonic spectrum.

The two extreme bands – at the far right- and left-hand side of the interface – govern the controls for the high-pass and low-pass filters within the EQ. Effectively, these completely remove frequencies

(Continued)

above or below the given cutoff point – try using the high-pass filter, for example, to tame any excessive low-frequency energy. Moving inward, the next two bands correspond to the shelving EQs, which are somewhat comparable to the treble and bass controls on a conventional hi-fi. Use the shelving EQ for general sweetening activities, creating the familiar "smiling" EQ curve with a boost at around 80 Hz and 12 kHz, respectively.

Figure 8.17 Channel EQ is one of Logic's primary mixing tools: with four fully parametric bands, two shelving controls, and two filter sections.

The remaining bands work as traditional parametric EQs, with a cut/boost control, frequency setting, and a fully variable Q parameter. Q sets the width of cut and boost, and therefore the resultant amount of cut and boost to harmonics is near to the EQ's selected frequency. A wide Q is a useful way of shaping more "general" qualities of the sound – maybe a lack of bite in the upper mids, for example, or an overall woolliness in the lower mids. Use a tighter Q where your frequency issues are more specific – like a boomy resonance on an acoustic guitar, for example, or prominent problem harmonics.

8.6 Adding Send Effects

As tempting as it is to patch everything across a channel's insert points, it's really worth making a clear distinction between the use of true insert-based effects and the use of send effects. The use of an insert effect implies there's no need to balance an unprocessed version and a processed version of the signal. EQ, therefore, is a perfect example of insert effect as there is little or no need to hear both an unequalized version and an equalized version of the sound at the same time. The same could also be said for compression. Reverb, on the other hand, is a clear candidate for being applied as a send effect, with the need to balance the respective levels of wet (in other words, reverberated) signal with the dry, unprocessed version. In theory, the greater the reverb, the more the channel appears to move toward the back of the mix.

The first step to creating a send from any given channel is to double-click in one of the available send slots on the strip and select from one of the 64 available bus destinations. Raising the level on the small send pot will then bleed an amount of the channel's signal via the bus to the appropriate aux channel strip. The aux channel strip should then have the corresponding plug-in (an instance of Space Designer, for example) patched across its insert path so that signals entering the aux channel strip are affected accordingly. With the mix parameter on the plug-in set to 100% wet, you can now control the level of reverb fed back into the mix, using the aux channel strip's fader or using further plug-ins across the aux channel strip to further process the sound of the reverb.

Both reverb and delay are important tools in defining the spatiality of your mix – in other words, the front-to-back perspective of the soundstage. Having worked carefully on your level and panning, you should have initially formed a good representation of your left-to-right soundstage. However, with the application of delays and reverb, you finally have a chance to set sounds forward and backward in the mix – a dry vocal, for example, sat squarely at the front, or a distant keyboard pad drifting toward the back of the soundstage. One really useful technique is to place two to three contrasting instances of Space Designer to establish a front, mid, and rear acoustic to in your mix. Use some of the shorter reverbs on close sounds, or rhythmic elements (like drums or rhythm guitar), that don't suit long reverb settings. The longer settings, on the other hand, could be reserved for a few unique sounds that really benefit for longer reverb tail and a more distant mix placement.

One important concept of the sends is whether they are configured for either pre- or postfade operation. In a mix, the most common example (and the one Logic defaults to) is to use the sends in postfade mode – that is, the send happens after any adjustment in level for the particular channel. If, for example, a fader level is brought down in the mix, its corresponding amount of reverb is also attenuated – in effect, preserving the ratio between the two sounds (dry and reverberated). Prefade sends (as we saw in Chapter 4) are usually associated in the creation of headphone or cue mix. To adjust among prefade, postfade, and postpan operations, simply click and hold on the send and adjust accordingly.

Figure 8.18 A basic configuration of send effects – the keys channel has been set up to have an auxiliary send delivering a little of its signal through to an auxiliary channel which is configured with a reverb on its 100% wet setting.

Figure 8.19 Use two or three complementary reverbs to help define the front-to-back perspective in your mix. Additional delays can also help define the "spatial" dimension.

8.7   Combined Processing Using Aux Channels

Having seen one application for aux channel strips – that of applying send effects – let's have a look at the other ways in which they can be applied during a mix-down. By changing the output option of any audio or instrument channel in the mix, you can also decide to route it to an aux channel strip ahead of the signal reaching any of the main, physical outputs from Logic. Used in this way, you can combine a number of signals – say, the various microphones positioned around a drum kit, for example – to a single fader, with the option to control both its level and the application of additional processing en masse.

The use of the so-called bus processing across aux channel strips can be a great way of locking together the principal components of a mix, as well as helping groups of sounds gel in an effective way. Many American rock engineers get a great deal of mileage from this technique, combining compression on a channel-by-channel basis, alongside compression across a collection of aux channel strips. The result is a mix that "pumps" in an empathetic way to the source material, providing a real intensity and sense of loudness brought about through the reduction of dynamic range.

However, if you do intend to apply a lot of compression on your mix, it's well worth making sure that you use a variety of different compression techniques as a means of forging the identity between different sounds (rather than the track just sounding squashed), as well as matching the style of compression to

Figure 8.20 Bus processing can be achieved by sending a number of channels (via a bus) to an aux channel. On the aux channel, insert the required equalization or compression, for example, so as to process the "group" of sounds.

the instrumentation you're processing. Try using low ratios (1.5:1 to 2:1) and low thresholds as a means of massaging the sounds into place, whereas harder ratios and higher thresholds (6:1, −5 dB) could be a great way of simply slicing a few loud transients of a drum submix. With any bus compression, you'll also need to pay close attention to the Attack and Release times of the compressor, with slower settings (especially on the release) providing a more musical result.

PLUG-IN BOXOUT 5
Convolution Reverb and Space Designer

Convolution has quickly become the accepted standard in producing professional-grade reverb treatments. The technique works by taking an acoustic sample, known as an impulse response or IR, from a given room or indeed, a hardware reverb unit. The process of convolution then simply takes the short IR file and mathematically "folds it over" the source signal, effectively recreating the effect as if the sound had been recorded in the same space or through the same reverb processor. For that reason, convolution reverb can sound astonishingly realistic, although this can be at the expense of your available DSP resources. Ultimately, therefore, it makes sense to run just a few instance of Space Designer, patched across bus faders, than lavish multiple instances on individual channel faders.

As part of the Logic Pro installation, Space Designer comes with an impressive Library of prerecorded IR files (stored under Library/Application Support/Apple/Impulse Responses). You can load these simply by scrolling through the presets or loading them in via the Load IR option to the right of the IR sample switch.

(Continued)

Additionally, you could choose to make use of one of the growing number of third-party IR libraries, like Spirit Canyon's Spectral Relativity, or indeed, your own custom-sampled IR files.

Figure 8.21 Based on unique samples of the original spaces, Space Designer's convolution reverb engine produces some stunningly realistic reverb treatments.

Like a growing number of the professional convolution reverbs, Space Designer offers a surprising amount of control and flexibility with effects you can achieve with it. Once an IR sample is loaded, you can immediately adjust both its volume envelope and its filter characteristics, assuming the filter is activated. Besides being able to produce some subtle modifications (rolling off the high-end, for example, with a touch of low-pass filtering), these tools can be great for abstract ambience treatments – try using the reverse setting and some extreme filter movements with lots of resonance! If you need to shorten the reverb tail, adjust the length parameter. Alternatively, to create a longer, darker reverb, adjust the sample rate into one of its slower settings. This can also be a great way of producing longer reverb times, without maxing out your DSP resources.

8.8   Using Groups

As alternative to bussing sounds together via aux channel strips, you can also make use of Logic's Group feature – as we've already introduced, with the concept of edit groups in Chapter 5. Taking the concept further, we can also use Groups to lock together collections of channels, say a group of drums, for example, or backing vocals. But given the aux channel strip bussing system, what use is an additional means of ganging faders?

As simple as the bussing system is, there's one major conceptual problem – that of postfade send effects like reverb. Imagine a collection of kit sounds all being sent to an aux channel strip, with a snare also making use of a generic reverb (also shared by the vocals and strings) on send 8. If the level of the kit were then reduced on the aux channel strip fader, you'd hope the reverb level would also be reduced accordingly, but this is not the case. As the send happens before the aux channel strip, any subsequent level changes will have no effect on the reverb – effectively the snare will get "wetter" as the level of the kit falls. Of course, one simple solution to this would be to route the reverb to the same aux channel strip, but this would then have a knock-on effect on the vocal and strings reverb!

So, here we see a perfect example of the application and benefit of grouping channels as apposed to bussing signal via aux channel strips. Grouping retains all the important gain structures of the original mix – the ratios of reverb are retained, even as the level of the group is reduced. Bussing still has its place, though, as there's no way to apply effects like compression and EQ onto the group. So, in theory, a good mix may well involve combination of both grouping and bussing via aux channel strips, grouping to "lock" collections of faders, and bussing via aux channel strips to provide grouped applications of effects!

Figure 8.22
Shared reverbs can make the use of aux channels for grouping purposes problematic; although the fader level of the drums group is reduced, the reverb level stays the same.

Figure 8.23
This example uses both groups and bus processing. The Group allows us to better control the level of the drums in the mix, while the aux channel is solely used for processing rather than level control.

Looking more closely at the Group Settings dialog, you can see a number of additional features that could potentially improve the speed and efficiency of your mix. Besides obvious grouping controls like volume and mute, you can also link the send controls (this could even be helpful in quickly modifying cue mixes as you're recording), as well as any linked Automation Mode selections. However, one vital keyboard shortcut to use in conjunction with the Groups is the Toggle Group Clutch command. By default, this is configured onto Shift + G and can be used to temporarily disable all groups – a great way of quickly modifying an "internal" balance within any number of groups, without having to laboriously enable or disable the relevant groups.

KNOWLEDGEBASE 2

Parallel Compression

Parallel compression has become a real buzzword in mixing circles, but what is it, and can Logic be realistically used to apply it? Unlike the traditional insert-based approach to applying compression, which works on the assumption that you'll only want to hear the 100% compressed signal, parallel compression offers the unique possibility of hearing both compressed and uncompressed versions. Although parallel compression is arguably less effective at reducing the dynamic range of a signal, it is an excellent way of combining both the hyped-up sound of heavy compression, with the natural dynamics and "air" of an uncompressed recording – yes, to use the cliché, it really is the best of both worlds!

One really simple way of creating parallel compression is to copy the source track onto two adjacent channels. On the one channel, keep the signal uncompressed and open, while on the other, try setting up really juicy, pumping compression. Don't worry about the integrity of the signal – pick a suitably tough ratio (6:1 or harder!), relatively quick Attack and Release, and let gain reduction pump well into the 6–10 dB range. Some engineers will even EQ this channel, picking out the extreme

Figure 8.24 In this example of parallel compression, a bus send is being used to add compression using the same routing technique as a reverb send. Use a heavy compression setting, and then bleed in the required amount compressed signal to add "body" without compromising transient detail.

(Continued)

highs and lows (80 Hz and 12 kHz), as well as applying a little tuck in the mids. With the compression set up, combine the two channels together, mixing in the compressed version to add "balls" and body to the uncompressed track.

Another interesting approach is to use the compressor almost like a send effect – blending any number of sounds through to the same heavy compression setting. Try doing this on a drum mix, sending mainly the close snare and kick drum through to the paralleled compression. Again, pick some suitably pumpy settings, maybe even a touch of extra ambience, and sit the bus in the mix to add the required amount of weight and importance.

8.9   Working with Channel Strip Settings

As you're building up the mix, you may well find that certain channels require similar settings to that of others – maybe you've compressed and equalized one backing vocal, and you want to apply the same setting to corresponding backing vocal track. Clicking on the small arrow to the right of the Insert label allows you to open up the channel strip settings menu. Try using the copy/paste option as a simple means of duplicating the channel settings on a number of channels. Alternatively, if you have a number of favorite channel strip settings – maybe a particular compression and EQ on a given vocalist, for example – you can also save the presets off, as you would any of the other plug-in presets. These favorites could then be recalled at any point on any given song.

8.10   Automation: The Basics

Despite the considered application of compression, you may still find the demands of your mix changing from verse to chorus and vice versa. Ultimately, although one balance may work at a given point in time of the song's development, the same might not be true as further instrumentation enters and the musical qualities of the track change. In the "golden days" of recording – before computers stepped out of the office and into the studio – it wasn't uncommon for a mix to involve "all-hands-on-deck": the engineer, assistant engineer, and even the members of the band, all frantically pushing faders up and down to shape the mix to the dynamic of the song.

Logic's Automation provides complete control over qualities of the mix throughout the duration of the song – including basics like volume, pan, and mute and more advanced options like plug-in parameter automation. The golden rule

Figure 8.25 Use the channel strip settings as a quick way of moving one channel's plug-in configuration to another. The presets feature also allows us to do this between songs.

to remember with automation is only to apply it toward the ends of a mix – once you've established the principal balance, equalization, compression, and effects usage, only then should you begin to turn to automation. The problem is that any mix changes after automation, although not impossible, can be quite a headache to perform – potentially requiring you to rewrite or delete existing automation moves.

8.11 Track-Based Versus Region-Based Automation

One potentially confusing aspect of automation is that Logic contains two contrasting methods for applying automation – track-based automation and region-based automation. In truth, region-based automation – where automation data is contained within the audio or MIDI region itself – is actually a throwback to the earlier versions of Logic and has largely been retained for the purpose of backward compatibility, although, to be fair, it still has its uses. Track-based automation – where automation data is recorded on separate multiple track lanes to that of the audio region – offers a far more flexible means of creating and editing automation data, as well has having the distinct operational benefit of not being tied to the regions in questions. Editing or moving a guitar solo, for example, won't necessarily disrupt the automation data that accompanies it.

Figure 8.26 Two different approaches to automation – the old-fashioned Hyper Draw (top) and the standard track-based automation (bottom).

Thankfully, Logic does provide a means of transporting one automation type (region-based or track-based) to another – making the combination of the two approaches quite a powerful feature. For example, you could write the automation using the superior track-based controls and then convert this information into automation data that is stored with the region. For now though, we'll take a look at how track-based automation works and then consider the region-based case later on.

PLUG-IN BOXOUT 6
PlatinumVerb

PlatinumVerb is a valuable alternative to Space Designer, either for Logic Express users (who don't have access to Space Designer) or for situations where you're running slightly low on available DSP resources. In contrast to Space Designer, PlatinumVerb places a surprisingly low drain on

(Continued)

the CPU, although correspondingly, you may find its output slightly less authentic than Space Designer. Used carefully however – on short settings with drums, for example, or any instrumentation sitting lower down in the mix – it can produce surprisingly effective results.

One of the best things about "modeled" reverb, as apposed to convolution reverb, is that it offers precise control over the sound produced in the virtual room. In the case of PlatinumVerb, it divides the reverb up into two key stages – the early reflections, as sound initially bounces back from the walls, and the more diffuse reverb tail, as the reflections merge together to create a definable "trailing-off" to the sound. Try moving the Balance slider over each side to fully understand and audition the effects obtained in each stage.

Figure 8.27 If you're short of processing resources, or don't have access to Space Designer, PlatinumVerb makes for an excellent alternative source of reverb.

Using a predominant mix of early reflections can be a great way of thickening up drum sounds by producing a noticeable, distracting reverb tails. Experimenting with the room shape, stereo base, and room size will change the qualities of reflections produced, either by creating a tight centralized reverb or by creating a wider, more expansive set of reflections. When it comes to the reverb tail, try exploring the density, diffusion, and reverb time parameters. On the whole, most conventional reverb effects tend to stay within the realms of 1–2 s, and, not surprisingly, this is where PlatinumVerb seems to sound at its best. Density and diffusion, respectively, govern the spacing and randomness of the taps. Try using both in their lower setting for an effect like Spring Reverb (great on electric guitars) or use a higher setting for a smoother sounding reverb.

8.12   Automation Modes

To engage a channel into automation, you'll need to change automation mode according to the way in which you intend to write automation data into Logic. By default, all faders are set to their Off position, just above the pan pot – this means that any existing automation data are ignored and the faders can be freely moved or repositioned (without fear of Logic snapping them back!) at any point. Engaging any of your faders into an automation Write mode (Write, Touch, or Latch) will allow you to begin writing data based on the current song position. Interestingly, though, the "recording" of automation data is independent of the transport's record switch; in other words, you only need to be in Play mode for automation to be written.

First, let's take a look at the various modes used to record automation data.

Write

Think of this mode as the most dangerous! Any fader engaged into write mode will record data onto the automation lanes, even if there's existing data, so this should really be used with care. Write, however, can be a valid way of deleting and replacing automation data in one pass.

Touch

Touch is the safer way of writing automation data, as it only engages into writing data once the fader has been "touched." Release the fader – even as the track is still playing – and the fader will return the previous recorded position and carry on reading any existing automation data. Touch, therefore, can be great in creating a few strategic nips and tucks in your mix – briefly either lifting a phrase out for a few seconds or pulling back any part that dominates the mix.

Latch

Latch mode is comparable to touch, in that a fader is only engaged into writing data once it has been touched. However, when the fader is released, it continues to write data – potentially erasing any existing moves – and the level it was last left at. Latch can be a useful mode when you need to raise or lower a level of a part and then leave it at that level for the remainder of the song without having to constantly "hold" the fader in place. Stopping the transport will, of course, stop the writing of automation data.

Read

Read is a "safe" automation mode, where any automation data is read back but no further moves can be written – indeed, if the fader is moved for any reason, it will promptly snap back into place! On the whole, most users tend to leave faders in Touch mode as they're automating, allowing them quickly engage the writing of automation data without having to constantly switch modes. However, to reduce the risk of overwriting certain aspects of the mix, it's best to switch all faders back to Read mode once you have finished automation.

Figure 8.28 Each of the four different automation modes used in Logic has its own impact on how you read and write automation data.

Although the description of different modes has assumed you're working with the channels faders, it's important to remember that once a channel has been placed into Write, Touch, or Latch, any movement in its accompanying parameters – including mute, pan, or any of the plug-in parameters – will also be recorded. This allows any part of the mix to be automated with the same degree of flexibility as the channel faders; for example, the reverb time of a vocal reverb could be swelled going into a "larger" chorus, or the feedback of a tape delay unit could be modulated to create some dub-inspired delay treatments on the end of a line.

PLUG-IN BOXOUT 7
Using Delay

Although most users tend to perceive reverb as the main tool for defining spatial qualities of a mix, it's also surprising to realize just how much spatial interest can be added with something as simple as a delay line. Logic comes with three different delay plug-ins – sample delay, tape delay, and stereo delay – facilitating a range of different delay treatments.

As the name suggests, tape delay is modeled on the type of delay effect produced by classic tape-based delay effects like Roland's Space Echo or the WEM CopyCat delay. As these processors use tape to produce their delays, the effect has a characteristic "dark and dirty" sound to it. By default, the plug-in works with tempo divisions – simply select the required division (semibreve, crotchet, quaver, and so on) and the delay will appear in time with your track, even if the song's tempo changes. Use the Groove slider in the extreme settings (33% and 75%) to change the division to dotted note values. You can also achieve some great slap-back delays (an effect famously used on early rock'n'roll vocals) by using a semiquaver setting and sliding the Groove parameter down into the region of 33%–50%.

Figure 8.29
Logic's simple delay plug-ins can be a surprisingly effective tool for mixing – from rhythmic delay effects to subtle forms of "slap-back" ambience.

Feedback sends a proportion of the sound back on itself, effectively creating a regenerative delay effect. On the tape delay, feedback settings in the region of 50%–100% will appear to "hold" the delay, with an increasing amount of distortion and grit on each repeat. Stereo delay works on the same principle as tape delay, although it deliberately avoids the tape-based coloration, in preference of a cleaner digital repeat. You can also set different delay time for the left- and right-hand side of the stereo image.

In contrast to the other delay plug-ins, sample delay only deals with incredibly small delay times, measured in samples. As a rough guide, 44 samples equate to 1 ms of delay, so even at its maximum setting (4000 samples), sample delay only gives us about 90 ms to play with! So why use this plug-in? Well, sample delay can be beneficial in applications fixing minute time delays caused when microphones are widely space apart – applying sample delay to correct these anomalies and create a more phase-coherent image. If you're less technically inclined, try using a small amount of sample delay on some drum room microphones to simulate the effect of sound reflecting further away from the kit.

8.13   Viewing and Editing Automation

By default, Logic hides the display of automation data – otherwise an arrangement could soon become cluttered. However, once you've started recording a few moves, you might want to see how the mix is beginning to shape up. Press "A" to display the current recorded automation. This is toggled so that you can quickly switch automation viewing on and off.

With the automation view engaged, you'll notice some important changes to both the tracks and the arrangement itself. Looking first at the tracks, you should now see the automation mode indicated on them (Read, Touch, and so on) alongside the current viewable automation parameter – clicking on this should allow you to scroll through all the parameters available for automation. You'll also see a small bar graph meter indicating the current fader position. Note that this can be freely modified and controlled, just like the "real" fader as part of the Mixer, and can be a great way of adding in a few cunning automation moves.

Figure 8.30
You'll need to enable the automation view mode to see or edit the moves you've recorded.

Where multiple plug-in parameters have been automated, you can also add further track lanes into the equation, simply by clicking the small arrow toward the bottom right-hand corner of the track name in the track list. Alternatively, option-clicking opens as many automation lanes as is required to display all the automation data currently recorded.

Automation data itself is displayed as a series of nodes, which can be manipulated using the usual Pen, Eraser, and Arrow tools. As you'd expect, the Pen tool allows to draw new automation moves, while the Arrow can modify existing node points, or by double-clicking add new nodes into the equation. Indeed, in many situations, it may be quicker and easier to draw in a couple of automation nodes than to laboriously write in a series of moves. Manually placing the nodes can also be a good way of setting in long automation events – a long filter sweep over 30 bars, for example – or when events, like fade-outs, need to happen at a precise time.

When two nodes are placed manually, you can also make use of the unique Automation Curve tool as a means of adding a degree or "curvature" to the line. Simply click and hold on the line between the nodes,

Figure 8.31 Press the small arrow on each track lane to open up a parallel automation lane, making it clear to see the moves on several different parameters on the same track.

Figure 8.32
Use the
Automation Curve
tool to create
neat and natural
curves between
automation
modes.

using the Automation Curve tool, and drag above or below the line, or from side-to-side, to create one of four different adjustable curve shapes.

Where ranges of moves need to be moved or duplicated, you can also drag-enclose a number of nodes simply by holding Shift as you rubber band a group of nodes. With the nodes selected, they can then be moved en masse, or with the option key held down, duplicated to a new position. One other potential lifesaver, in case you've written in the automation moves but feel the need for them to be a couple of decibels higher or lower, is Logic's scaling feature. To scale a track's automation moves, use the Command key and drag up or down on the small bar graph meter as part of the track list. Look carefully at the selected values, and you should be able to spot the corresponding changes you've made.

PLUG-IN BOXOUT 8
Vocal Processing

Vocals can be one of the trickiest parts of a mix to get right, so it's great to see a range of plug-ins suitable for vocal processing in Logic. Away from the core plug-ins that we've already covered – including EQ, compression, and reverb – you'll also find the DeEsser and Pitch Correction plug-ins helpful in crafting that polished vocal performance your mix requires.

Logic's DeEsser is a handy way of taming any excessive sibilance (like the letters s and t) that might have been accentuated by a poor choice of microphone. The DeEsser works by analyzing the input and then applying a selective amount of gain reduction whenever problematic sibilance is heard. The Detector part of the interface is what should be used to spectrally locate the sibilance in the vocal. Set the Monitor mode to Det (detection) to tune into the specific frequency your signer's sibilance is occurring (usually somewhere in the region of 6 kHz). Now, flick to the Sens (sensitivity) monitor setting and adjust the sensitivity so that it only flicks on when the sibilance occurs. Moving the monitor to its OFF position, adjust the Suppressor parameter to the same parameter to that of the Detector and increase the strength to get the required amount of sibilance reduction.

Figure 8.33
Logic's DeEsser and Pitch Correction plug-ins can aid a number of problems in relation to your vocals.

The Pitch Correction plug-in is loosely based on the infamous Antares AutoTune plug-in and provides an on-the-fly means of correcting intonation problems in a vocal – or indeed, any other monophonic – performance. At extremes, using the response setting on its 122 ms setting, the plug-in can be forced to produce the clichéd, quantized vocal effect so carelessly abused in the late 1990s! On softer sensitivity settings, however, it can be a useful way of taming any problematic pitch drifts. To achieve the best results with the plug-in, you'll need to specify the key and scale the song is in. A number of presets for this are available, but you can also manually switch the notes in and out by clicking on the accompanying keyboard.

8.14 The Automation Menu Options

Besides the various graphic tools for editing automation data, there's also an accompanying automation menu by right clicking (or Control + Clicking) containing various options to delete moves, as well as some intriguing options to move data back and forth between track-based automation and region-based automation (as previously discussed). Although it is, of course, easy enough to delete selective parts of automation using the eraser tool, the menu options make quick-and-easy to create "blank-slate" on individual automation lines, tracks, or indeed the whole song itself. Certainly, in situations where you might inherit a previous song file for a new project, this can be a great way of clearing out problematic automation moves.

Figure 8.34 The Track Automation menu includes several powerful features, including the ability to move automation between its native track-based form and the region-based solution of Hyper Draw.

Moving track-based automation to region-based automation is useful in situations where you'd like the automation permanently attached to a part, without having to continually specify for the automation data to be moved every time you realign or duplicate the part. Possibly the best "real-world" application of this would have to be the use of automation to control filter movements – maybe you've written a couple of distinctive filter movements that are as much part of the musicality in the region as the notes contained within the MIDI sequence. By switching the automation over to region-based (Ctrl + Click and then Always Move Automation With Region), you effectively lock the movements into the part, allowing yourself to forget about whether the data are moved or not. Of course, at any point, this same data can be brought back to track-based automation by Control Clicking and returning the feature to Always Ask whether to move the automation or not.

KNOWLEDGEBASE 3
Saving CPU: Freezing and More ...
At some point in the mixing process, it's highly likely that you'll start to hit dreaded inevitability of a CPU overload. Dealing effectively with this problem could make the difference between a half-baked mix and a distinctive mix, using appropriate strategies to save CPU resources without compromising on sound quality. First, it helps if you're clear as to which plug-ins are the real CPU

(Continued)

monsters in your session. For example, synthesizers like the ES2 and Sculpture can take up a big quota of resources, while the EXS24 (assuming it not making lots of use of its filter) is surprisingly processor efficient. In respect to plug-ins, Space Designer can be really CPU-hungry, especially if you're using longer IR files in excess of 2 s, while the Channel EQ is almost negligible on today's Intel-powered machines.

Arguably the first technique, therefore, is to look for any optimizations you can apply to the session that might improve overall CPU efficiency. For example, try turning off unused oscillators in the ES2, or removing excessive Space Designer use on inserts, in preference for a few instances on aux sends.

The next step is to use the track freeze option. Freezing effectively creates an audio bounce of the track in question, complete with all its inherent plug-in settings, and then deactivates the plug-ins accordingly. If you then try to edit any part of the track – either opening an instrument or plug-in or repositioning a region – Logic will remind you of its "frozen" status. If you do need to carry out the edits, simply unfreeze the track, make the modifications, and then refreeze.

Freezing itself is carried out with the small freeze icon as part of the track header. If you can't see the icon, select Alt+T and then add the freeze button into the set of options. Note that the freeze occurs across the full duration of the project, so it's worth moving back the project end marker to the real finish point of your track.

As an alternative, you can also carry out an audio bounce yourself, simply rendering complex virtual instruments as audio regions, saving off the instrument settings, and packing away the MIDI data (just in case you need to go back). This can be an effective solution earlier in the project where you intend to do lots of structural rearrangement with the regions at a later point.

Figure 8.35
Use the Freeze function to render an off-line version of the track and so release valuable CPU resources.

KNOWLEDGEBASE 4
Compression Circuit Types

The qualities of the different compressors are something that tends to excite many professional engineers, mainly as the different approaches to circuit design can achieve some radical differences in the types of compression achieved. Ultimately, this means that certain compressors tend to suit themselves to specific applications, bringing plenty of "character," as well as gain control, to the input they're processing. In the example of Logic's compressor plug-in, therefore, we're provided with Platinum, ClassA_R, ClassA_U, VCA, FET, and Opto.

Although the exact models from which these circuit types are derived aren't supplied, it is worth noting some of the key differences between these different options and how best to apply them. Opto represents the oldest compressor design, based on early models that used optical cells as part of their gain control circuitry. This unique design resulted in a degree of latency – both with respect to the Attack and Release on the compressor – that tends to deliver a more "musical" compression, as apposed to a harder, more aggressive gain control. The Opto circuit type, therefore, works well with bass sounds (that don't suit fast Attack and Release times), vocals, or anything that you want to retain a degree of musicality and lightness with.

Figure 8.36 The different circuit types approximate the unique sonic behavior of many classic types of compressor like the Urei 1176 or LA-2A.

The FET design, on the other hand, used Field Effect Transistors to create a more heavy-handed compression, particularly good at catching transients. The FET circuit model therefore, and to our ears the ClassA_U, can produce some really effective results on drums, especially when used across overheads. In this application, don't be afraid to use low threshold setting with "pumping" Attack and Release times for an aggressive, almost low-fi compression sound.

Although this provides a theoretical background to select the circuit type, the best approach is to use your ears. Try configuring some basic compression settings and then flick between the different compressor models to hear the marked differences in how they sound.

WALKTHROUGH
Adding Compression
Step 1:

Insert a compressor across the instrument you want to process and start by establishing some basic settings. Working from the default positions, try finding a ratio and threshold setting that works for the instrument you're trying to process. For example, for a gentle compression, use a medium threshold with soft ratio (1.5:1 through to 2:1) yielding about 2–3 dB of gain reduction (another term for compression). For a harder compression effect, consider bringing up the ratio (4:1 or more) and lowering the threshold to achieve 6 dB or more of gain reduction.

Figure 8.37 Basic compression setting.

Step 2:

With these basic settings established, you can now start to refine the compression a little. Try adapting the Attack and Release settings to best suit the style of compression you want and the sound you're trying to squash. Slower settings (Attack 40 ms, Release 400 ms) tend to create a more natural progression in and out of gain reduction, although you might find the occasional loud transient slipping through the net. Faster Attack and Release settings (Attack 0–10 ms, Release 100 ms) produce a "pumping" effect, which tends to work well where you want the compression to sound more noticeable.

Figure 8.38 Refining the attack and release.

Step 3:

As compression leads to an overall loss in level, you need to raise the output to restore the overall signal to its original peak level (although, of course, the signal will be more compressed). Try bypassing the compressor, noting the meter readings, and then using the Gain parameter (with the compressor active again) to restore the original level. Listen carefully to the compressed sound in the mix – is there enough compression to sit the instrument correctly? Is the compression too obvious? Further fine-tuning (maybe increasing the ratio or softening the Attack and Release) optimizes the compression for the instrument's position in the mix.

Figure 8.39 Increasing output gain.

LOGIC TIPS

Plug-In Delay Compensation

Any plug-in added into a channels signal path creates a small amount of delay or latency, through the extra processing required in producing the effect. Thankfully, however, Logic includes a feature to compensate for any delays incurred through plug-in processing, called plug-in delay compensation (PDC), available under the general tab of the audio preferences (Logic Pro X > Preferences > Audio). For users using standard audio unit plug-ins (either Logic's own or from other developers), stick to the "audio tracks and instruments setting." However, if you are running a processing-accelerator system like Universal Audio's UAD-2system, you may notice delays building up when you start to use buses to apply UAD-2 plug-ins. In these situations, change the preference to its All setting – Logic should now play the entirety of the session in time.

Figure 8.40 Changing the PDC settings is important for users of processing-accelerator systems such as Universal Audio's UAD or TC Electronics' PowerCore.

PLUG-IN BOXOUT 9

Pedalboard

Pedalboard is an ideal companion plug-in to Amp Designer and is designed to replicate the array of footpedals that guitarists might use to shape the sound of their guitar. As such, you'll probably want to place Pedalboard ahead of an instance of Amp Designer, although of course, you're free to use it anywhere along your signal path should you see fit. Indeed, the combination of both Pedalboard and Amp Designer shouldn't just be reserved for guitars, as the two plug-ins can work wonders on a range of sounds – from synth basses, to low-fi breakbeats, and vocals!

(Continued)

Pedalboard is easy to use – simply drag the required pedals from the Pedal Browser on the right-hand side of the interface over to the pedal area found on the left-hand side of the plug-in. The signal path works from left to right, with each subsequent pedal adding the previous pedal's output. This means the order of the pedals can have a big effect on the eventual output, although it's generally best to place "tone" pedals, like compressors, equalizers, and distortion pedals, earlier on in the signal path; with effects pedals, including flange, chorus, delay, and spring reverb later on in chain of footpedals.

As well as using a left-to-right "serial" routing, Pedalboard also allows you to create parallel routing configurations where you can split the signal between two separate busses and then sum the results at the end of the chain of footpedals. For example, this makes it possible to layer two different types of distortion, rather than each distortion unit adding to the results of the previous pedal, or to send a split a sound to be processed by reverb and delay as separate entities.

You can assign a pedal to Bus B using the Routing area above Pedalboard – simply click on the pedal you want to move, and it switches to the upper Bus. Note a Mixer object now appears at the end of the signal path, allowing you to blend and pan the two discrete signal paths (A and B). Note that you can also manually insert a Splitter pedal to achieve similar result, although the added benefit here is that the Splitter pedal can also divide the sound by frequency, allowing each bus to process a different part of the frequency spectrum.

Figure 8.41
Pedalboard replicates the array of footpedals guitarists might use ahead of their amp to shape the basic sound of the guitar.

In this chapter

Knowledgebases

Plug-in focus

Mastering in Logic

9.1 Introduction

If you've ever tried burning a few tracks onto a CD, you'll be well aware of the challenges of producing a CD that sounds comparable to a commercial release. Even with a complete mastery of the production process, and some great-sounding final mixes, your CD could still sound weak and, comparatively, amateurish. Although OS X and iTunes offer integral CD creation and burning, the fact is that professional musicians and bands will invest a significant amount of money and experience in turning their finished mixes into a final product. So does this mean that users of Logic can't enjoy the same degree of finesse and polish? Well, with a little know-how, and the audio tools of Logic, you too can produce release-quality mastered audio.

The term mastering describes the process of compiling and editing several (possibly contrasting) recordings, applying some form of audio "sweetening," and assembling these to produce a final Red Book "production master" CD. Traditionally, mastering has necessitated the use of a separate facility (other than the music studio), specially equipped with mastering equipment perhaps including multiband compressors and high-end mastering EQ, and dedicated workstations like Sonic Solutions, Pyramix, or Sequoia. Nowadays, however, the world of media is a lot more demanding, so it's not uncommon for musicians, bands, and composers to master by themselves. The bar, it appears, has been raised – and Logic can meet the sonic challenges!

9.2 Different Approaches to Mastering

Mastering itself can be divided into several objectives. First, of course, is the sweetening we most commonly associate with a commercial CD – in other words, the use of compression, EQ, and a host of other processes across the finished two-track master. Second, tracks will also need to be edited – setting correct start and end points for example-or placing any fade-ins or fade-outs as required. Finally, the finished files or regions need to be ordered for the CD, with appropriate markers to define the tracks and index points that appear on the CD. The first two objectives are easily possible within Logic, but the final stage may require you to look at other solutions to make a Red Book CD ready master. We'll now look at three techniques of getting your music onto a CD now.

Technique 1: iTunes

The first technique is the idea of applying mix sweetening, edits, fade-outs, and dithering all options inside the main Logic application. The finished files are then rendered (using Bounce to Disk or Logic Pro > File > Share discussed later) and exported as 16-bit 44.1 kHz files ready for compilation elsewhere.

The compilation process, however, requires the use of other software – either iTunes (Apple's integral audio-CD burning tool) or dedicated Red Book software like Jam (see knowledgebase on external Red Book standard CD software).

Figure 9.1 Audio Masters, exported from Logic, can be assembled into a finished CD using iTunes or any other suitable Red Book standard software.

Technique 2: Burn from Logic

For the quickest and the most integral solution to the mastering problem, you can burn CDs directly from Logic. Again, sweetening and edits can be applied directly to the main application. The burning of CDs, however, is carried out from the Bounce menu – rather than rendering an audio file, you'll burn the finished mix onto a CD. This method is, arguably, the quickest way of creating a CD, especially with just one or two tracks needing to be burned, but it doesn't offer the most flexible solution in the long run.

Technique 3: Export to Another Application

The best technique is to employ a separate Red Book standard CD application. Apple has stopped the development of Waveburner, originally part of the Logic Studio package, but dropped for version X.

Figure 9.2 For a quick CD, use Logic's integral CD Burn feature, as part of the Bounce to Disk option. Although this method is an easy way of burning single tracks (a rough mix, for example), it can be too restrictive for professional mastering.

Alternatives are available such as Roxio Jam and Wavelab to mention just a few. The reason for employing these is the important Red Book options like International Standard Recording Codes (ISRC) and CD-Text that are not possible directly from Logic X.

9.3 Bounce to Disk

Whether you are mastering in Logic itself or using another application, the first step will be to render your Logic mix as an audio file. The temptation to apply audio sweetening or fades in the main Logic project can be persuasive, but this should be avoided at all costs. Ideally, if you have at least one copy of your song unmastered at 24 bits, then it would be more suitable to take to a professional mastering engineer (when the lucrative record deal arrives!) than a home-mastered, 16-bit file. Separate high-resolution files also afford you the opportunity to approach mastering your work away from the "headspace" of a mix, so that tracks sound correct in terms of the whole CD rather than on an individual basis.

The Bounce dialog window can be accessed through the Bounce, "Bnce" button, on the main output channel strip. Before clicking on this, you'll need to define the length of the bounce, designated by the current cycle length – without this, Logic will simply default to bouncing from the beginning of the first region to the end of the last. By defining the length manually, you can keep the best account of a few crucial factors – namely, the "hangover" at the end of the track generated by reverb tails and the allowance of a small amount of silence at the start of the audio file. At this stage, it's probably best to render a slightly larger file, rather than too small, as trying to add information later (say, for example, when a reverb tail gets cut off) can be tricky, if not impossible.

In the Bounce window, you'll need to specify the file type, resolution, and dithering options. Ideally, a PCM, AIFF or WAV, 24 bit, 44.1 kHz, interleaved file is considered the best and current "raw data" for mastering, unless you've recorded at a higher sample rate than 44.1 kHz. More important, you should check that the dithering has been set to none, as dithering is best applied at the very last stage of mastering when the word length is reduced to a 16-bit master. The bounce itself can be carried out in real time (maybe you've got some live synths or effects coming into the audio mixer) or offline – a quicker way of rendering to file by temporarily devoting all your computer's resources to the bounce process.

Figure 9.3 When performing a bounce in your Logic arrangement, check that the duration of the bounce is slightly longer than your track and the resolution is 24 bit.

KNOWLEDGEBASE 1

Pre-mastering

Officially, the term "mastering" refers to the process of cutting a master disc from which the records would be pressed in duplication. Pre-mastering originally described the process that engineers used to prepare the audio signal for the vinyl medium. Nowadays, the term mastering tends to be used in place of pre-mastering and describes the process of preparing the audio for the intended medium, ordering the music and also signal processing.

Logic's Share Menu

In Logic X, there is a new menu called Share (Logic Pro > File > Share) that contains three options. The first is to export directly to iTunes to make that all important playlist of your tunes. It can be good to compare your masters here to other professional releases using the shuffle mode. You'll soon see if your track sticks out or not.

Figure 9.4

An alternative way to distribute your music would be by using the Share feature in Logic and exporting to the media browser for use with either an external CD application or iTunes. Note the export to Soundcloud integration here.

Figure 9.5

Sharing with iTunes will reveal a window permitting you the ability to populate the basic fields about the tune such as the Artist, Composer, and Track Name.

The second option permits you to add the shared audio file into Logic's Media Browser (and therefore Finder also). This feature is good if you're at the end of your mix and wishing to bounce ready for mastering. Alternatively, you could be at the end of mastering and just wish to email the audio to a client for approval.

The final option is sharing to Soundcloud integration. This will prove popular with Logic users who are increasingly using this medium as a way of expressing their audio output.

Figure 9.6
Under the Share to Media Browser option, the filename is the only field available for alteration, and usefully Logic places in the time and date for you. It is then up to you to provide a name. Notice the options for file quality also.

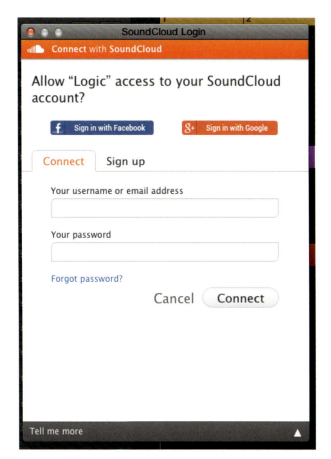

Figure 9.7
The Share to Soundcloud option initially opens up a log in page where you can log in using your Soundcloud details, Google log in, or even Facebook. Once logged in, you're permitted to populate a number of fields and specify the quality of the audio and whether it can be streamed or downloaded.

9.4   Audio Mastering in Logic

With the raw data of your tracks assembled, you can begin to look at mix sweetening. With the various songs pulled into a master Logic session, you could place each region on a different track and experiment with various plug-ins to reach the desired sound. Ideally, the CD should present a rounded and uniform tone throughout, with a consistent "loudness" across the tracks – you could even try importing some commercial tracks as a reference to see just how far you can take things. Exactly how you achieve this will vary from track to track.

The principal tools used in mastering are well represented in Logic, with a number of dedicated mastering plug-ins, including phase-linear EQ, multiband compression, and limiting. Obviously, you can use any plug-in where appropriate, but these particular tools will be the most useful in achieving a professional sound. Plug-ins can be inserted either on the individual track's insert points (for song-specific processing) or across the main master-channel strip for general application (maybe when the album needs to be limited as a whole, for example). As with any audio processing, the order of the plug-ins is vital for the end result, although the widely accepted order for mastering is EQ, followed by compression, and finally limiting. Metering (applied across the main output using the Channel EQ's Spectrum Analyzer or the Multimeter plug-in) will help keep an overview of things – note, in particular, how the commercial tracks might meter differently to your own mixes.

9.5   Editing Fades

Besides preparing the sound of the tracks, it's also important to look at some other important mastering details, like the exact start and end points, and any desired fade-ins or fade-outs. Having already mastered audio editing in Chapter 5, you should have no problems switching over to working on whole songs rather than individual regions. When you're editing the start point, however, make sure you leave a small amount of silence and don't cut right to the beat – this allows for older CD players to demute and stop the track from "jumping-out" at the start of playback. Also, pay close attention to unwanted noise at either end of the track – the easiest way to address this is to either use the Sample Editor's Functions menu to "Silence" unwanted noise or use a quick fade-in and fade-out as appropriate.

Creating proper track fade-outs (i.e., fading out over the last chorus) is a little more taxing. The problem lies in the fact that adding a fade using Logic's conventional method – arrangement area fade tool – doesn't account for the use of mastering processors. As the fades are pre-effects, any corresponding change in the level of the track will result in a change in compression – so as the track fades out, for example, it slowly loosens its compression. To counteract this, try performing another bounce, this time with the processing in place. The new "compressed" region can then be reinserted into Logic (the old effect being disabled), and then the fade can be applied. An alternative method would be to automate your fader of your stereo track as this is post-effects.

Figure 9.8 Mastering in Logic using a combination of channel inserts and the main stereo output's insert points.

Figure 9.9 Use Logic's Sample Editor to perform basic "top and tailing" tasks, like silencing unwanted noise or creating small fades (as illustrated).

Figure 9.10
For longer fade-outs, you'll need to render the file with effects and then create the fade using either Logic's region fade tool (as illustrated) or the Sample Editor's fade.

PLUG-IN FOCUS 1
Linear-Phase EQ

Having looked at Logic's standard channel EQ in the mixing chapter, let's take a look at the mastering-orientated Linear-Phase EQ. Being much more CPU intensive than the Channel EQ, the Linear-Phase EQ produces a technically superior sound by removing the phase shifts (caused as frequencies are cut and boosted) that occur in a conventional EQ. Even with its CPU drain, the Linear-Phase EQ is a welcome and sonically accurate tool for enhancing and shaping the timbre of a track in mastering. Operationally speaking, the Linear-Phase EQ includes the same controls as found on the Channel EQ and the same phenomenally useful FFT analyzer.

Applying EQ in mastering requires an approach different to that of mixing. On the whole, your approach needs to be as subtle as possible – remember everything you do will be much more noticeable as you're only working with two stereo tracks! Try to keep boosts, or preferably cuts, to a maximum of ±3 dB with a wide bandwidth (low Q parameter) – if you're going beyond this, you may well have a problem with the original mix. Overall, your track should exhibit a smooth response, with a gentle roll-off of high-frequency energy. Remember though that there are other tools that can also have a timbral effect on the mix – sometimes more successful than EQ – like multiband compression (especially on bass) or harmonic excitement (for some top-end sparkle).

Figure 9.11
The Linear-Phase EQ removes any of the usual phase shifts that occur with conventional parametric equalization.

9.6 Exporting and Burning

Once the tracks have been processed and edited correctly, you can begin to create the final exported 16-bit masters ready to be burned, or indeed, you could burn the files directly from Logic. The Bounce dialog window again comes into action – either to create the final rendered files or to provide access to the Burn feature. Given the regions of the exact finished song length, you can use the Region menu to select "Set Locators by Regions," and in this way, the cycle length (and therefore the bounce) will be exactly the same length as the region. The Bounce dialog window should specify the creation of 16-bit files, which will necessitate the (final) application of Dither to smooth out any quantizing noise brought about by moving away from 24 bits. Once exported, these files can be dragged straight into iTunes (or other suitable software) ready to be burned.

With burning now supported directly from the Bounce dialog window, you can also burn a CD directly from Logic itself. Again, where the session contains audio files greater than 16-bit resolution, the addition of dithering is essential, which can be selected from the appropriate pop-down menu in the Bounce window. Other pertinent options include setting the write speed (keep this low to avoid any write errors that degrade the audio quality on playback) and writing the disc to a multisession CD. When the "Write as multi-session" option is selected, you'll be able to add further mixes to the CD at a later point; otherwise Logic will simply burn a single file onto the CD. Although adding sessions at a later point will save the use of CD-Rs, you might experience problems with CD players recognizing songs burned in subsequent sessions.

In many ways, the "Bounce and Burn" feature is best viewed as a quick fix for creating audio CDs – especially if you intend to write more than one track to the CD at any given time. In this scenario, it might be,

Figure 9.12 Using the "Set Locators" feature in the contextual menu entitled Playback, you can set the locators to your selected regions, enabling you to create a bounce exactly the same length as that of your chosen audio.

perfectly, legitimate to attempt some rough mastering in the actual project's file, rather than a separate session – simply insert a Multipressor and some EQ across the main stereo outs and away you go!

Figure 9.13
The last bounce will need to be dithered down to 16-bit resolution, ready to be burned in another application.

Figure 9.14
Quick and easy CD creation: Logic's Bounce and Burn feature.

KNOWLEDGEBASE 3
POW-r

POW-r comes with three options within Logic. One is labeled "Dithering" that employs a noise-shaping curve to reduce noise. The remaining settings offer two types of noise shaping. The Logic Manual describes the first noise-shaping algorithm, which can "extend the dynamic range by 5–10 dB." The second noise-shaping algorithm is intended for work with speech as it can "extend the dynamic range by 20 dB within the 2–4 kHz range – the range the human ear is most sensitive to" (Apple Logic Manual). These noise-shaping curves have been developed to be sympathetic to the human ear and draw upon considerable research beginning with Fletcher and Munson's Equal Loudness Contours (more information on Equal Loudness Contours can be found in Rumsey and McCormick (2002)[1]).

9.7 Creating a Red Book Master

For a fully professional CD that meets the requirements of Red Book standards, you'll need to use an external application. Preparing for this will simply involve rendering your files (probably at 24-bit, without dither) ready to be imported into a mastering or CD compilation application. If you need to separate out your mastering "headspace" to that of you as the producer, then Logic can work, but it must be remembered that it is a dedicated production tool. Track crossfades, for example, are difficult to create in Logic but is part and parcel of mastering applications. Additionally, mastering applications offer support for CD-Text, UPC/EAN barcodes, and index markers. See the knowledgebase for Red Book CD mastering applications for further exploration.

KNOWLEDGEBASE 4
Red Book Standard CD Applications for the Mac

Apple used to provide a good application for creating Red Book standard CDs called Waveburner, which was part of the Logic Studio suite up until version 9.0 (not later downloadable versions of 9). This is now a discontinued product and will have new Logic users looking perhaps for alternatives. The good news is that the development that went into those specialist plug-ins are still available as standard in Logic Pro X. As Waveburner is not available now, we've taken a little time to signpost you to a number of alternatives depending on what you want to do and your wallet's size currently.

If you're looking to master within a new Logic session and then simply compile the audio and produce a DDP, a favorite is Sonoris' DDP Creator (www.sonorissoftware.com). It is a professional-level compiler for the mastering engineer wishing to place audio together and create a DDP image for CD production.

For the software to prepare a Red Book standard CD with some audio editing and sweetening features, you might wish to look at the following. A relatively affordable alternative is Audio File's new Triumph Software, which replaces their former Wave Editor software. www.audiofile-engineering. com/triumph. It has some excellent features for audio editing, contains some iZotope code inside

(Continued)

for mastering, and is fully Red Book CD and DDP compliant. The downside is that it is quite tricky to get the hang of, using a proprietary method of "layers" for arranging its audio.

A simpler option and possibly the closest in terms of operation to Waveburner is DSP-Quattro – www .dsp-quattro.com. Even the panes are in similar places, and the software does very similar things for editing and arranging your CD. Where this software falls down is on the plug-ins. If you've got something like iZotope's Ozone 5 or other VST or Audio Unit plug-ins for the job of mastering, then this is a great, DDP-enabled application. Note the Logic plug-ins don't appear to work in the version we used.

Figure 9.15 Possibly the nearest thing to a replacement for Waveburner is DSP-Quattro.

Others may choose to explore Steinberg's Wavelab that is certainly looking toward a fully fledged mastering platform, but is pretty expensive. There's also cheaper Wavelab Elements that can achieve most things but do not provide DDP support. See www.steinberg.net for more details.

Above these are the more professional alternatives such as Sequoia http://pro.magix.com or Pyramix (www.merging.com).

PLUG-IN FOCUS 2
Multipressor and Multiband Compression

Originally developed as part of Emagic's and later Apple's, now discontinued, Waveburner software, Multipressor lives on as Logic's answer to the big multiband compressors (like the TC M6000 or TC Finalizer, for example) so frequently used in mastering. Unlike conventional single-band compression, a Multiband compressor splits an incoming audio signal into separate frequency

(Continued)

bands before applying compression. Separated in this way, a complete mix is far easier to control and gel to form a finished master. In most cases, the greatest amount of compression will be applied to the bottom end of the mix – keeping the bass solid and tight – with the high-end requiring a light and minimal touch.

Multipressor features up to four bands of compression and downward expansion, although, to make things easier, you might just want to use three bands (low, mid, and high, respectively). One of the most important things is to set the right crossover frequencies for the different bands, as this can have a big effect on the finished compression achieved. To set the crossover point, try moving the vertical borders on the top of the interface's window – you could also use the Spectrum Analyzer (part of the Multimeter plug-in) to visually analyze the track's constituent components, as well as the band solo feature.

Now, work your way through each band, adjusting the relative compression ratios, thresholds, and attack and release settings. On the whole, most compression ratios tend to fall way below 3:1, with

Figure 9.16 The Multipressor allows you to compress your master using a series of different frequency bands.

(Continued)

the most productive results often achieved in the area between 1.5:1 and 2:1. Adjust your threshold to achieve the correct amount of gain reduction and dynamic control. Again, mastering is often distinguished by a lightness of touch, with typical amounts of gain reduction rarely exceeding 2–4 dB. Attack and release times change the responsiveness of the compressor – avoid setting the release too fast (the compressor might start to pump) or squashing the transients too much with a fast attack. Finally, rebalance the bands using the Gain Make-up control. If you've compressed a band particularly hard, you might need to bring its level up to restore its position back into the mix.

PLUG-IN FOCUS 3
Limiting
Although Multipressor does a good job of bringing the overall dynamic range of your track more in-line with a commercial CD, it's still possible that your tracks will sound at least 4–6 dB quieter than a professional release. This is because fast transients (often from slightly louder drum hits) have slipped through Multipressor's "net" and are still dictating the overall peak level of the track – even though the average level is somewhat below this. Professional mastering engineers will, therefore, make use of a fast-acting digital limiter (sometimes known as a Brickwall Limiter) to catch these peaks and reduce them. With the peaks reduced, the overall track level can be lifted to match the loud output currently favored in mastering.

Figure 9.17
The Adaptive Limiter plug-in provides that extra sense of loudness, although you need to be careful on how hard you push this addictive processor.

Logic's Limiter presents a basic set of controls that should allow you to squash transients in no time. The main parameter, gain, pushes the overall mix harder toward 0 dB – where loud signals would then conventionally distort logic outputs, the Limiter kicks-in and applies gain reduction (indicated

(Continued)

in the top meter) to stop this from happening. In most cases, especially those that have already been through the Multipressor, you'll only need to apply about 3–4 dB of gain, although you will need to listen carefully for any unwanted audio anomalies, especially distortion.

PLUG-IN FOCUS 4
Other Mastering Tools

Besides the three big mastering tools (EQ, multiband compression, and limiting), Logic also features other plug-ins suitable for track sweetening. Stereo Spread, an effect often found on hardware-mastering processors like TC Electronics' Finalizer, offers the ability to extend the width of a stereo signal – say, for example, the mix had been made slightly too mono. Unlike certain other Stereo Spreaders, Logic's mix widener avoids the use of phase in achieving extra width – instead, it uses a frequency distribution system to pan alternating frequencies to the left- and right-hand speaker, respectively. By using this frequency distribution method, Stereo Spread avoids the phase problems often associated when a width enhancer is put back into mono. In truth, its effectiveness is somewhat limited, and it only works best with material possessing little or no stereo information.

Figure 9.18
Although they have little everyday use, Logic's other mastering plug-ins can be useful solutions to a range of mastering problems.

Denoiser uses some clever trickery with FFT filtering to produce a cleaner audio signal in situations where the master possesses a large amount of unwanted noise. Like a conventional noise gate, Denoiser needs its Threshold set carefully to make the best use of the effect – try locating a quiet section of the master, with the noise present, and set the Threshold just above this. Reduction defines the amount of Denoising taking place, although if pushed too hard the Denoising can sound almost as distracting as the noise itself! The Noise Type fader seems a little strange at first, but it does make sense – in the center position, the fader is sensitive to noise across the entire audio spectrum, toward the top its bias is toward darker noise, and to the bottom it is more sensitive to high-frequency noise.

KNOWLEDGEBASE 5

Disc Description Protocol

Logic is unable to write an image to the Disc Description Protocol (DDP). This protocol, developed by Doug Carson and Associates, is a way of describing the whole audio CD as simply data. Until recently, only top flight mastering applications allowed this. There are now a handful of Mac applications that can enable you to complete this task – see the knowledgebase for Mastering Apps.

DDP images can be written to a CD-R, or a DVD-R, dependent on file size, but can be placed in a folder for sending over the Internet. DDP offers a degree of quality control that is not always available from a simple burned audio CD. As the DDP is essentially data, it follows high-standard error correction, and as such, it is unlikely to work if it is in the slightest bit corrupted. However, if it works, it should be perfect, and this ensures that the quality of your master reaches the pressing plant in perfect condition.

If you're hoping to get any number of CDs pressed up in the traditional way using a glass master, then DDP is the file protocol for you.

For further information on mastering, consider our book *Practical Mastering: A Guide to Mastering in the Modern Studio*, also from Focal Press – www.masteringcourse.com

Note

1. Rumsey, F., & McCormick, T. (2002). *Sound and Recording, an Introduction*. Focal Press Oxford, UK.

In this chapter

Walkthrough

Knowledgebase

Logic and Multimedia Production

10.1 Introduction

Writing music to picture for film and television once required a whole host of equipment from some form of video player synchronized to your computer sequencer and Digital Audio Workstation. However, Logic takes this task in its stride, and today, whole soundtracks can be written for picture from one application. This has been made possible by the thorough integration of a visual player and high-level synchronization options from the ground up. This flexibility has meant that Logic has become one of the most popular products for film and television music production.

Additionally, in the past few years, we have seen a sudden rise in other outlets for the music producer, such as gaming and Internet multimedia. Logic, with its integration of codecs to export audio as MP3, AAC, and many other formats, has positioned itself as a key production tool for many industries other than mainstream recording and production.

In this chapter, we take a movie file and begin integrating it into the Logic project and explore how to write effectively for film and television. We'll look at working with surround sound, markers, scoring, and the delivery formats expected from the industry.

10.2 Managing Movies

Logic has been at the forefront of music and audio composition for picture for sometime, and as such, there are many ways you can work with visuals. You can, of course, synchronize Logic to an external Visual Editor such as Avid or Final Cut if you so wish, but you're more likely to obtain your video as an .avi or .mov file with embedded timecode so that you can work wholly within Logic itself.

Figure 10.1
To get started on your composition, you'll need to select Open Movie from the File menu.

Before starting work on the music and audio for your visuals, you will need to import the movie into Logic. To begin to integrate movies within your song, go to Open Movie (File > Movie > Open Movie …).

Once you have located your movie in the browser, it will appear as a floating window, which sits neatly on top of Logic. This can be moved around to fit the screen arrangement you have currently. This floating movie can be resized to suit, and this is achieved by Ctrl + clicking the video window. This produces a new small selection box that reveals itself to offer you the choice of an alternative size for the window or the Video Project Settings.

Figure 10.2 Ctrl + clicking the movie window opens up many options such as the size of the movie window itself.

The Video Project Settings allow you to control a number of things starting with where the movie presents itself in the timeline.

You can specify how loud the volume is from the video and whether to mute it. In other words, the audio from the original movie can be piped through to the main output, although you may wish to control this in more detail by importing the audio into the Logic project, which we'll cover later. Other flexibilities are available here, such as the movie start. For example, it is often unlikely that you'll want the video to start running from 00:00:00:00, but may indeed need some preroll for any introductory credits that might not have come to you yet.

Figure 10.3 The small movie viewer in the Inspector can be a real space saver when working on a smaller screen.

Additionally, you can specify that the movie follows the tempo, although it is suggested that you don't automatically entertain this.

To the bottom right of the Video Project Settings menu, there is a "Movie Preferences …" button. Clicking this reveals a few settings starting with the Movie to Project slider which alters the start point of the movie in relation to the timeline. This level of fine-tuning can be very useful when lip syncing or making overall macro adjustments.

Next is the Movie Track section which first shows the Cache Resolution drop-down menu. This allows you to set the resolution of the thumbnails used in the track header. Following this is the Maximum Cache Size slider which defines the very maximum amount of video that can be held in memory. The manual offers some guidance of how to use this, but again we'd recommend that these settings are left alone in the main.

Finally, the Lock Movie Window When Changing Screensets essentially maintains the video screen's position and size floating above each of the screensets.

If you're working on a small monitor or a laptop, Logic has a neat way of monitoring what is going on in your movie. Close the floating movie window and at the top of the Inspector, there will now be a movie window in the inspector bar. Simply click on the triangle on the left-hand side to open up a small movie viewer.

Extracting the Movie Audio Content

Quite often, you might be given a visual that has integrated audio, perhaps a dialog track for the scene you are composing to. It will be prudent to extract this dialog track for you to be able to manage its level and mute throughout the writing process. In some circumstances, for example, you may wish to mute the dialog to ensure that you can hear your music properly before reintroducing it to see if it all fits together.

The quickest way to extract audio for this purpose is to select Import Audio from Movie (File > Movie > Import Audio from Movie). By selecting this, Logic goes to work immediately taking the audio from within the movie and moving the subsequent audio file to a track on the Logic Pro Main Window. The other option is to import audio from movie, which places the audio file in the Project Audio Browser. When the audio is extracted, its region will be timelocked to the position of the movie to ensure lip synchronization.

Figure 10.4 Importing guide audio from the movie can make embellishment a little easier.

10.3   Global Tracks

Global Tracks are an important feature of working to picture, not only for the viewing of video stills in the timeline but also for the ability to change the tempo, time signature, and keep track of important markers in the project. Changing the pace of the music can be crucial when composing to picture, and Global Tracks allow you to make these adjustments effectively as required.

Accessing Global Tracks is achieved by pressing "G" while on the Logic Pro Main Window. This should expand to show you any number of lanes that can typically include markers, signature, and tempo. To add to these, we need to choose to configure Global Tracks.

For example, a common way to view the visuals you're working with is to use Global Tracks. Video stills can be added to this timeline by going to pressing Alt + G or control + clicking in the Global Tracks area (see Fig.10.5). A list will present itself in which you can select the lanes you wish to see in Global Tracks. By selecting video, a new lane appears which can show a set of stills of your video against the timeline.

Within this lane, click on the Movie word in the header and revealed are some useful buttons. One of those is Open Movie, which offers you the opportunity to load a movie to your Logic arrangement. However, if you've not yet loaded a movie to the project, then a Finder-like pane will appear to allow you to select a new movie. The other button is Create Marker Set from Scene Cuts, which we'll discuss in a moment.

Figure 10.5
Right clicking in the Global Tracks space opens up a useful menu and at the bottom is the Configure Global Tracks.

Figure 10.6
Clicking on the word Movie, or any other Global Track header title, provides a contextual menu with useful features.

Markers

As with the locators found on traditional multitrack tape machines, Logic can also offer the ability to return to any marked position on the timeline. This can be useful to return to any point of a project which is visited frequently when recording, such as the beginning of the Middle 8 or a Chorus. Additionally, this can have some excellent uses when working to picture.

Logic terms these location points as markers, and they can be seen in the time bar at the top of the Tracks Area but are accessed in two main ways, through either the Global Tracks or the Markers Tab in the List Editors pane. As you expand the Global Tracks, you will notice that the Markers Lane come into view. It is likely that this lane also needs to be expanded by pressing the triangle to the left of the track header. This lane shows any markers you may have set against the timeline.

To get instantly working with markers, simply place the playhead at the required position and then click the Create button. A new marker is created that appears to run straight to the end of the project, although there is no length specified. This will be interrupted by the next marker you create or if you specify a marker length.

Markers can also be created on the fly while you are playing. This can be useful to note any aspect of a recording or a take you wish to revisit when you listen back. By pressing Cmd + Alt + Ctrl and clicking on

Figure 10.7 Scene markers are clearly seen within the Markers Tab and can be renamed and edited here.

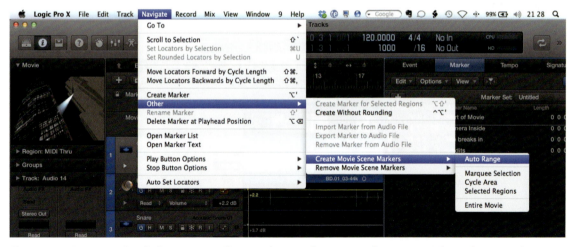

Figure 10.8 For more detailed extraction of scene changes for your markers, you might wish to use the menu at Options > Marker > Create Movie Scene Markers … where there are various selections to make scene cut detection more tailorable.

the Lane, a new marker is created, whose name awaits you to edit. Simply rename the marker and press return. This can be really flexible as you label sections going through a song, or film, for example.

Editing a marker can take place directly on the Global Track Lane by selecting the appropriate tool from the Click Tools Menu ("T") and then editing the marker. Using the Text Tool to name makers is very flexible and can assist your workflow while you play through the takes.

All markers are also reflected on the Marker tab within the Lists Pane, which can be found by pressing Alt + M while on the Logic Pro Main Window. Alternatively click on the Lists button to the top right of the Logic Pro Main Window. Next, select the Marker tab that shows all markers within your project in a long editable list. Here, you'll see the markers arranged in a table with their position to the left-hand side and their Length to the right. Naming may appear the same as in the Global Tracks lane here but can be added to create a separate line within the Markers Lane, which can reflect the part of the project. Simply double-click the name of the marker within the tab to reveal the marker naming area.

Each marker is tagged as "Marker ##" or "Scene ##," where the "##" automatically numbers the markers in sequence. If you continue to write after this, then your text will be merged on the same line. But, as in the example below, by pressing return, the Marker Lane can produce separate lane to show your description, which can be very useful when working to picture.

Scene markers are different to that of traditional markers in that they are locked to the movie file using Society of Motion Picture and Television Engineers (SMPTE) timecode. They will not alter depending on the tempo changes or edits in the project. It is possible to convert these to normal markers and vice versa by selecting the Marker tab (Alt + M) and selecting the appropriate Convert … command from the Options menu.

Locating to the marker positions is simple and can be achieved first within the Marker tab by using the Finger Tool from the Click Tools Menu (usually the alt + Click Tool is set to this automatically) to click on any marker. The playhead will spring back to the marker point. To wrap the locators around the marker length, press Cmd + U. To access the start of a marker from the Global Tracks lane, simply press Option + Click a marker. Key commands can be programmed to respond to marker numbers if required, but are not part of the default key commands set.

Within the Global Tracks Marker Lane, click on the word Marker to reveal a menu with "Create Markers From Regions" in it. This feature allows you to select pertinent regions from which markers are to be made. Perhaps you have different folder tracks already cut to the length of your verse and chorus. In this instance, the markers would automatically follow suit and allow you quick access to these elements of the song.

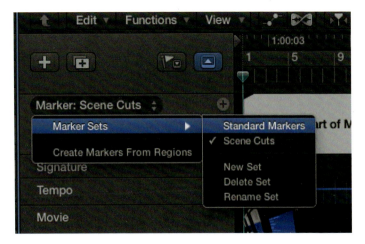

Figure 10.9
Alternative marker sets are available from the global tracks header Marker menu.

Alternative sets of markers can be selected within the Global Tracks Marker Lane Menu. This offers the opportunity to have up to nine different sets of markers. This might be very useful when working to picture where markers are required which represent the dialog, others can be set in an alternative for working against the music and perhaps another for the visuals themselves.

Detect Scene Changes

It is important to get a feel for the structure of the visuals you are working with. A script or storyboard may not be provided and yet you will need to break the visuals down into useful, manageable chunks. The best way to get a handle on the visual material you are working with is to work scene by scene.

"Add Scene Cuts to Marker Set" is a button on the Movie Global Track and stands for Detect Scene Cuts. Logic has a really useful feature in that it can detect scene changes in the video and produce markers for you to refer to or edit to suit. These take a little while to detect and will appear in the Marker Lane in Global Tracks and in the Markers List press "D" then click on the Markers Tab.

Scene markers are presented with a "film" icon within the list and are automatically labeled as Scene 1, Scene 2, Scene 3, etc. Obviously, this process may not be entirely 100% accurate, and it may be necessary to alter, add, or delete some of these Scene Markers. To create extra markers to denote scenes that the process may have missed, simply create a new marker by pressing the "+" create marker button on the marker Global Track. This marker may need to be converted to a Scene marker as described above by opening the Markers Tab in the Lists area, then opening Options > Convert to Scene marker.

As you work through each of the Scene markers, you will need to move them to fit the movie more accurately. This can be easily achieved in the Global Track, where you are automatically given the resize tool. Alternatively, visit the Marker tab and edit using the SMPTE timecode.

Working with Tempo and Signature Changes

Making changes to the pace of the music is often the key to writing effective music when writing to visuals. Whether it is to denote a new mood or some dramatic news that has been announced within the dialog, changing key and tempo, in conjunction with the accompanying visuals, could instantly move your audience to another emotional space.

Changing the tempo is very simple and can be achieved in the Global Tracks. Either select the pencil tool (T > 2) and click the tempo change at the desired place or choose the Tempo tab from the Lists pane, which can be accessed by the icon on the right-hand side of the Logic Pro Main Window.

Using the Pencil tool can offer you a quick method to try ideas out in the project. Clicking the Global Track will reveal a new tempo, which can be glided up or down to instigate a tempo change. You'll notice that the way in which these tempo changes are represented look very similar to that of the automation lanes and as such can be manipulated in similar ways.

For example, you might choose to introduce the piece at a slower tempo and increase it gently. To achieve this is easy, simply set the two different tempos at the ideal point. The jump will be instantaneous and

Figure 10.10 Tempo changes can be altered quickly and flexibly in the lists pane on the Logic Pro Main Window or by editing the Tempo Lane within the Global Tracks.

will sound odd. There is a small point in blue at which the tempo changes from the original tempo to the new tempo. This first point can be moved backward in time to introduce a more subtle curve to the tempo change, thus making it gradual.

The menus within the Tempo tab offer some more options to tempo management. For example, you can view the subtle steps that Logic has placed in the project to increase the tempo. To do this, you need to click on the Additional Info button to the left of the Edit menu in the Tempo tab. Prior to selecting this feature, only the start and end point of the tempo change can be seen.

With the steps expanded, it is possible to quickly edit any step to suit. As you edit the tempo of one of these discrete steps, the Global Track will reveal the steps graphically on screen. However, to create a more staggered, creative, or editable tempo change, simply choose Edit > Tempo > Tempo Operations, which can offer some more options to the on screen Global Track.

A new pane appears that allows you to edit more precise and in-depth tempo changes and alterations. There are many different editable features here, such as the types of curves and their resolution. Within the Operation menu, there are a number of really useful features that could save time, such as stretching the curve out or simply scaling it a little.

Figure 10.11 The Tempo Operations dialog gives some additional control over making tempo alterations.

The signature lane in the Global Tracks is split into two; the top is for the time signature and the other for the key. As with the Tempo lane, there are again two main methods of changing the signatures. The first is through the lists on the right-hand side Lists window, which contain a tab dedicated to signatures. Within this tab, there are two buttons, one for Time Signatures (create signature) and the other to Create Key changes.

The second entry method is to click again on the lane itself with the Pencil tool (or simply double-click with the pointer). Whether it is for a signature or a key change, a slightly different dialog box greets you. The first of these is the time signature that not only allows you to simply select the signature you want but also gives you the opportunity to enable beat mapping, which we covered in Chapter 5.

Adding a key change is really useful for quick ideas and changes and will work with Apple Loops and any audio you have recorded in this Logic song. The key signature here will enable the Loops to alter

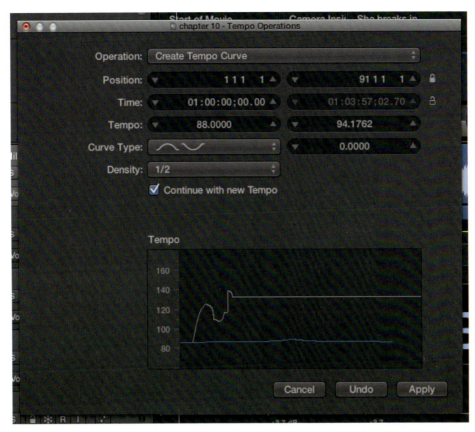

Figure 10.12 The Tempo Operations dialog gives some additional control over making tempo alterations.

Figure 10.13 Adding a key change in the Global Tracks is easy by double-clicking in the lane where a dialog box allows you to specify the change.

in real time to your changes. Simply select the point at which the key change is supposed to occur and double-click (or click one using the Pencil tool), and a dialog box will emerge requesting the desired key and also whether you wish to disable double flats and sharps.

Alternative Sets

It is often difficult to be sure that the choices you make are definitely going to be right for the outcome of the visuals. Often, it is nice to make changes and try out ideas without losing the original version. Logic has an interesting solution that allows different experiments.

Each Global Track has something called Sets, which offer different perspectives on the same track. Similar in nature to alternative playlists when quick swipe comping which we covered in Chapter 5, the Sets menus in the Global Tracks allow you to make some different decisions without losing your original root idea. Simply click on the Sets menu for each of the Global Track lanes and choose an alternative number.

Figure 10.14
Global Track Sets are useful to try out different permutations of ideas in Global Tracks.

Big Displays

Working to picture is always dependent on timecode, which governs the synchronization between different cameras and audio recorders for editing purposes. Timecode typically comes in the form of SMPTE, which splits the time into divisions of Hours:Minutes:Seconds:Frames, and then in some cases, subframes.

The Control Bar usually always shows the SMPTE time of the playhead in the top left with the accompanying bars and beats below depending on the view. When spotting sounds or effects to film, it will probably be necessary to see a larger, and independent, timecode display. Ctrl-clicking the Control Bar's icon on the left-hand side (see Figure 10.15) brings up a menu from which you can choose from the Bar Display or SMPTE Display. These are displays that remain within the darker section of the Control Bar. There are also two options to show these as separate floating windows by selecting the Open Giant Bar Display and Open Giant SMPTE Display, respectively (shown on next page).

Figure 10.15 In Logic Pro X, the Transport Bar has been renamed the Control Bar and has a slightly different way of working, such as accessing the menu from the icon to the left-hand side of the bar.

Spotting Audio

Working to picture will from time to time mean that you will be given a spotting sheet from your clients listing the sound effects. This will tell you where all the sounds need to occur to a particular timecode point and may also include an out time point. These points are given in SMPTE timecode, and in Logic, it is easy to edit to these points. It's worth sometimes working to a snap value of "Frames" when working to picture to ensure that each frame is synchronized with the audio.

Two main ways of spotting audio are offered. First is by using the Move to Playhead Position feature. Simply set the playhead to the point you wish for the audio to be placed. This can be best achieved by double-clicking on the SMPTE display in the Control Bar and typing is straight in. Pointing toward the audio region, click with control pressed down to reveal the contextual menu and within this will be the Move > Move to Playhead Position command.

The second and more likely method is to edit using the Event tab from the Lists window on the right-hand side of the Tracks Area. Presuming you have called up all your audio regions onto the Tracks Area, they should be listed here. To hone in on the parts you wish to spot, ensure that Link is on both the Events List and the Logic Pro Main Window, and simply click on the region you wish to view. The Events List Tab is naturally set to show bars and beats and will need to be changed to view SMPTE through the View Menu (View > Event Position and Length as Time).

Figure 10.16 The Customize Control Bar dialog box provides considerable flexibility in the information and control you have over your project.

Figure 10.17 The playhead's SMPTE position can be specified by double-clicking the timecode readout in the Control Bar.

To spot the audio, simply double-click on the position column to type in the desired timecode. The audio should in theory be spotted, although it might be likely that you will need to delve in further to more accurately ensure synchronization with the picture. Nudging the regions by frames can be a really quick way of making things fit. As we mentioned before, it is often wise to work to frames for the time being, although there will be times where you wish to work in finer resolutions. To change the Nudge Value, right-click on any region and select Move > Set Nudge Value To. Once set, simply select the region to be moved and select either the Nudge Left or the Nudge Right from the region's contextual menu or press Alt + Cursor Left or Alt + Cursor Right.

A really fast way to spot audio is to position the playhead at the desired point and select the track you wish the audio to be placed. Next, open the separate Project Audio Browser Window using Cmd + 8 or Window > Open Project Audio if you wish for a separate window view. Select the audio file in question by pressing Cmd as you double-click, and the audio file will be placed at the appropriate place. If you then select another audio file in this way, it will be butted up to the last audio file. There are a couple of useful key commands that can be brought in here called Pickup Clock and move Event to Playhead Position (key command ";" and Pickup Clock and Select Next Event (shift + ";").

10.4 Synchronizing Logic

Working to picture or on larger projects might involve the need to link Logic to another player. For movie work in the past, it had been often the case that the video would be on another machine, perhaps tape-based, and Logic would need to synchronize to it. Similarly, there may be times when Logic needs to synchronize to another multitrack system, whether that be a legacy open reel 2-inch 24-track tape machine such as a Studer A800 and Otari MTR90 or simply another ProTools rig on another computer. Either way, Logic has a suite of features to cope with your synchronization needs.

Logic syncs in a number of different ways that are industry standard. The main system used for video is SMPTE, and it is an audible code that can also be translated into a digital equivalent used in MIDI called MIDI timecode (MTC). MTC is a popular system used between MIDI devices, and in the latter example mentioned earlier, it is likely that this would be the protocol used to connect a ProTools rig to your Mac with Logic on.

In either case, you will require a device that understands the timecode protocol. All MIDI interfaces should understand MTC, but only some devices are able to translate audible SMPTE timecode to MTC such as MOTU's MIDI Express XT.

Figure 10.18 The Sync button seen here in blue must be selected from the Customize Control Bar options to appear. With this selected, Logic works in slave mode to incoming synchronization signal.

Figure 10.19 Right-clicking the Sync button on the Control Bar gives you very quick access to the pertinent settings such as synchronization source and settings.

Getting Locked Up

To tell Logic that it will need to slave to another timecode signal requires you to select the Customize Control Bar preferences by Ctrl + clicking the Control Bar. Select the Sync check box that will generate a button on the Control Bar, which looks like a clock with a large arrow facing into it.

In an ideal world, the synchronization should be as simple as selecting this button, and wait for an incoming sync signal to get the project moving. However, invariably you will need to edit the synchronization settings to suit the project you're working on. To do this, either Ctrl + click the Sync button on the Control Bar and select synchronization settings or vising File > Project Settings > Synchronization.

Within this pane lie a number of key features for working with external synchronization sources. The frame rate of the synchronization is very important. Historically, frame rates differ depending on where in the world you work; for example, 30 frames per second is used for television in the United States of America, whereas we work with 25 frames per second in the United Kingdom. This is a simple division on the frequency of the electricity supplies in these regions at 60 and 50 Hz, respectively.

Strictly speaking, we're no longer governed in quite the same way, and Logic can work at whatever frame rate is required. Within this menu are considerably more frame rates than we've hinted to above and it will all depend on the project you're working on as to what frame rate you'll need. Safe to say, Logic can handle them all!

Within the synchronization settings pane are some opportunities to offset the timecode to the arrangement. These offsets are very important when working with film. For example, you may wish to use a fresh Logic project for a Scene that is one hour into the film. It would be rather silly to start the Logic project at 1 hour also because things like Freeze tracks would take a considerable amount of time to process!

Figure 10.20 Use the synchronization settings to select frame rates and the usual offsets needed from time to time.

Setting the offset is easy. Simply select the SMPTE time at which your song should play. For example, Logic will open with Bar 1 playing at 01:00:00:00:00, which means 1 hour. SMPTE reads as Hours:Minutes:Seconds:Frames:Sub-Frames. However, there will come a time when you will need to alter this to suit. In the example above where we'd like to start the new Logic project at 1 hour into the movie, we'd need to set an offset of an extra hour. As such, we'd want the SMPTE reading here to read 02:00:00:00:00. Logic will then offset the incoming timecode to allow Logic to start playing at 2 hours.

Below the main offset selection is an option to allow the bar positions to show the absolute timecode reference, which can be useful when working to external sources. It is important, for example, that the timeline you work to reads the same as the incoming timecode. Due to the offset in the example above, the Logic project will show a time of 00:00:00:00:00 despite being an hour into the film. This can be changed here, so the project starts Bar 1 at 1 hour, but also shows an accurate portrayal of the timecode it is receiving.

Figure 10.21 The frame rate will need to be matched to your synchronization source to keep up!

Being the Master

Logic also has the ability to be the master and to allow other devices to slave to its timecode. In this instance, it is necessary to return to the synchronization settings as outlined above and click on the MIDI tab. Here, we can see a checkbox titled Transmit MIDI Clock, which refers to the tempo settings of the project. This clock pulse can be outputted using the MIDI protocol to connect other devices such as effects units whose delay settings can respond in time with the project. Alternatively, in rare occasions, this can be used in part for synchronization.

Below this is the Transmit MTC checkbox and output menu. Here, it is possible to select a device (or "All") to send the MTC to. This will enable any device connected using MTC to follow Logic. For example, if the ProTools rig was to sync as a slave to Logic, then simply connect a MIDI cable between the output of Logic's interface to the other computer with ProTools' MIDI In.

MIDI Machine Control is a remote control protocol that enables connected equipment to be controlled by another device. For example, Play on the master could be engaged by controls on a remote slaved device. This can still be a hugely beneficial option to employ across large studios.

Figure 10.22 Transmitting MTC can be engaged by selecting an output from the Transmit MTC menu, or leave it to send out of all available MIDI outputs.

10.5 Score Editing and Music Preparation

Although the Score editor can be used for MIDI editing activities in Logic, its most useful features lie in the "preparation" of music – in other words, taking the raw MIDI information that you've performed in the project and transcribing this as finished, musical notation. To be fair, though, Logic doesn't compete with the publishing-standard output of dedicated industry-standard scoring programs like Finale or Sibelius, but its features are more than adequate either to produce parts for a small-scale overdub session, for example, or in the case of a full orchestral session, a sensible "intermediate" format to present to a proper orchestrator.

The process that we're going to explore here, therefore, is the rudiments of taking an existing MIDI composition and turning that both into a full score for the conductor and parts for the musicians. Rather than being a complete exploration of the scoring features and the art of orchestration (which is a book in its own right!), we're going to take a look at the essential processes and steps that guarantee readable, usable results in the shortest amount of time. If you want to produce even more effective scores, though, it's well worth exploring further to see just how effective the Score editor can be in this task.

Preparing Your MIDI Files

As the old adage goes – garbage in, garbage out – this is never more true than in the process of creating a score. Although it's easy enough to open any region in the Score editor so as to see a notation-based view of the music, it doesn't guarantee that the score is legible or playable. Arguably, the clearest example of this is strings. In your MIDI arrangement, it's highly likely that the entirety of the violins, violas, and cellos, for example, will be amassed to one generic string patch. Although this polyphonic approach makes sense for MIDI production, it isn't how players expect to read a score, as in reality, each instrumentalist will be expected to be presented with a single line, with the full score combining all the single lines onto a page.

Figure 10.23 Here's a typical "massed" string part, with all the violins, violas, and cellos being triggered from the same general MIDI track.

The first step to prepare an effective score is to divide the music into a number of distinct lines (one for each player, or part) within your Logic arrangement. As we saw in Chapter 6, there are a number of ways of doing this – from simply copying the part over and deleting notes in turn to features like Functions > Note Events > Assign MIDI Channels based on Score Split, followed by Edit > Separate MIDI Events > by Event Channel. Either way, you should end up with a series of regions for each part, each of which should be assigned to a unique named track for each instrument you want to appear in your score (like violin I, violin II, viola, and cello).

Figure 10.24 Creating a score from a series of split parts, each assigned to a different named track, will produce a more legible overall score, as well as facilitating the printing of individual parts.

With the parts split, you might also want to glance through and double check any timing issues, as well as the precise duration of notes. For example, in a MIDI arrangement, it's easy to have notes finishing 1/16th before the end of a bar, but this can lead to some strange looking note durations in the score, alongside unwanted rests. For simplicity, it's also worth merging each part into a single region lasting the full length of the score, unless you want parts dropping in and out of the finished notation.

Opening the Score Editor

With the MIDI data prepared, you can now go and open the required parts in the Score editor, by drag-enclosing the required instruments and opening the Score editor via the Window menu. Although, of course, you can also open the Score editor directly from the Logic Pro Main Window, it probably makes most sense to have the complete screen dominated by the scoring features.

On opening the Score editor, you should have something on screen that approximates music, although as with the raw MIDI data, this will need to be tweaked to make the notation as legible as possible. For example, most of the instruments will probably default to a piano stave (with both a treble clef and bass clef on the same line!), and it's probably the case that the music simply runs across the full width of the screen, rather than being displayed on a page-by-page basis.

Figure 10.25 On opening the Score editor, you'll be presented with a basic score, although this will need to be modified to make readable.

The first change, therefore, should be to turn the Score editor from its standard view into a page view, via View > View Mode > Page View (ctrl + P). Next comes the task of setting the staves. In keeping with the workflow on the Logic Pro Main Window, you'll find that each line of score has its own set of score parameters, available via the Inspector to the left-hand side of the Score editor window (note that this might be minimized at first or even with the Inspector itself hidden). By clicking on each line of the score, therefore, you should be able to select an appropriate style for each part – maybe selecting treble for the violins or the bass clef for a cello part. Note that for any transposing instruments (like Horns in F), Logic will automatically transpose the "scored" version of the instrument, although the "played" version remains untouched.

Adding Expression and Score Markings

To add interest to the score, it is vitally important to add expression markings; otherwise, you end up with a sterile and lifeless recording. To help with this, the Score editor provides a part box, which allows you to place a series of expression markings onto your score. These could be as simple as pp, mf, or ff dynamic markings, right up to trills, or crescendos and diminuendos. What you should notice is that the score stylings are attached to a particular part or line in question (the corresponding part will be highlighted blue when you drop the object) so that even when we move back down to part level (rather than the full score), these markings will be carried with the part.

Figure 10.26 Adding expression markings, and so on, will make your score both clearer and more expressive.

Alongside expression marks, the part box also allows you to set the key of the piece (drag this to the first bar of the score and then any subsequent bar should a key change occur), as well as text-based objects like the song name.

Shrinking to Fit: Score Settings and Score Sets

Changing the broader qualities and appearance of the score can be done via the score tab of the project's settings, also assessable via the Layout menu. Although most of these settings (margin spacing and so on) will be best left in their default state, there are a number of useful functions. For example, the Number and Names tab allows you to change how instrument names are displayed, using either their Full Names throughout the entire score, for example, or an abbreviated Short Name for pages after the first page.

Figure 10.27 The Score tab under the project settings allows you to define some macro properties about how the score is laid out and printed.

Another part of the full score's appearance is the percentage scaling of the staves on the page. For example, on a full score, it's often important to see as much of the music as possible, so as to avoid excessive page turning, as well visualizing the entirety of the arrangement at any point in time. This is of particular relevance to a full orchestral score, where you might need 30–40 different staves on the page at any one time.

Technically, when we selected our particular group of regions, we created what is known as a score set. As the name suggests, a score set assembles a group of instruments that will appear together on the score at the same time, theoretically letting you omit certain tracks on the Arrange pane from being displayed in the finished score. In addition to this, you can also define a percentage scale of a score set, allowing you to shrink or expand the score's size accordingly. Go to Layout > Score Sets … to access the Score Sets Window, which you'll need to ensure has its "local" Inspector open in the window so as to access the Scale [%] parameter.

Printing the Parts

To print off individual parts from the score, try double-clicking on the stave in question. This should drop you down one level (the equivalent of going into a folder in the Arrange pane), allowing you to

Figure 10.28 Use the Score Sets Window to alter the overall percentage scaling of the score.

see a single part with its associated score markings. So as to make the part clear and legible, you might want to consider raising the Scale [%] parameter as part of the Score Sets controls. Remember, though, that with both parts and the full score, you have the option to print to a PDF (via the OS) so as to create a savable "hard" copy of the score for future reference, printing, or indeed, as an e-mail attachment for your music copyist.

10.6   Surround Sound in Logic

Surround sound, while synonymous with the film industry offering that extra dimension of realism, is now becoming popular in some circles for music production with many artists' back catalogs being mixed for the medium. Whether working to picture or in music production, you'll inevitably find yourself being expected to create a surround sound mix at some point. Fortunately, Logic has a great deal of expertise in offering surround sound with the inclusion of surround sound panners on each channel and a 5.1 Reverb plug-in among other features.

To work with surround sound, in this example 5.1, it is important to ensure that you have the speakers you require to monitor each output. You'll need your main stereo pair of monitors, plus a center speaker, two surrounds (left and right) plus a subwoofer ideally. Connecting these up will usually mean a direct connection from your audio interface or desk to the amplifiers or powered monitors. The connections you make are governed by the Surround Preferences (Logic Pro > Preferences > Audio > I/O Assignments > Output). Within this pane are three tabs relating to the Output, Bounce Extensions, and Input.

The input arrangement should follow logically the output assignments and is used for making 5.1 recordings. The bounce extensions tab simply refers to the filename extension which is important when preparing for mastering. Concentrating on the Output tab, you can select how Logic shows its outputs. Typically, this will default to Logic's own interpretation of 5.1. There are options for International Telecommunications Union's (ITU) and WG-4's output arrangements. The ITU's is standard for 5.1 surround sound for most professionals. However, the WG-4 standard, which can be selected by the button below, is the choice of the DVD forum. At this stage, it is just important that the right outputs are connected to the correct speakers for accurate monitoring.

Getting Started

To get started with surround sound in Logic, you could either select a surround sound template from the chooser or alter your current stereo production to work in surround. Choosing Logic's own template offers 24 surround-ready channel strips all bussed to groups including a 5.1 incarnation of Space Designer. This is excellent for that 5.1 project you're about to start.

However, there will be times when you'll need to alter your session to output in surround. To do this, simply choose the channel strips you wish to pan in surround and then visit the output assignment button. The drop-down menu should now show the possible output assignments, the last of which is Surround. Click this and notice the master channel strip alter from stereo containing two peak meters to a surround fader with 5 meters.

Figure 10.29 Selecting which audio outputs from your interface are sent to which speakers can be an important part of setting up your surround sound studio.

Working with Surround is very similar operationally to that of stereo except panning is different. You will have noticed a circular panner on each channel. When double-clicked, this opens up the Surround Balancer pane that shows a good-sized panner showing a visual representation of the space with the five speakers facing inward. To pan a mono signal, simply drag around the ball within the space to inform Logic where you'd like it to come from. Logic will calculate how much signal should come from each speaker.

For stereo signals, Logic's Surround Balancer shows three balls. The "L" and "R" balls represent the two discrete mono signals that go to make up the stereo signal. These balls are linked and allow you to consider how wide you want the stereo signal to go within the surround space. The third ball is the overall pan that will alter the direction the stereo image comes from. In the larger surround panner on p. 373, the stereo signal is shown biased to the right.

Other controls on the Surround Balancer include the center level and the low-frequency effect (LFE) level. The center level refers to the image created by the center speaker. As surround sound is derived from the film industry, the center speaker has traditionally been placed behind or around the projection screen and due to its focal point it contained all the dialog. As such, the music was usually omitted out of this speaker, leaving it to create the phantom center image it currently creates with stereo reproduction. With a center level control in Logic, it is possible to decide whether to allow any signal to come through the center speaker at all.

Panning stereo signals in surround can make the process of mixing somewhat complicated. With two speakers, it is clear where the stereo signals go and how they might be managed as they can only come from within the space provided between the two speakers.

Surround sound in its popular incarnation means that there are five speakers through which the stereo signal can be routed. Does this stereo signal simply go to the left- and right-hand speakers and faithfully represent itself, or is it reinterpreted for surround and maneuvered around the sound space? In the past, a stereo pair on a guitar might have stayed centralized to the center, but with a multichannel set up, it is possible to say rotate this 90° so that the left-hand channel could be reproduced from the right monitor and the original right channel from the rear right monitor, hence keeping the stereo image but shifting it in the sound space as shown in the example below.

Figure 10.30
The Output Assign menu on each channel offers you the ability to choose any available physical mono or stereo outputs, in addition to a surround mode.

Figure 10.31
Clicking and holding on the button with a circle(s) in it will offer you the opportunity to quickly select the output mode of the selected strip.

Figures 10.32 and 10.33 In this view, the panner can be seen managing a stereo signal. When panning stereo sources, there are extra controls to change the separation of the signals.

Four further controls are available from the bottom of the Surround Balancer. To obtain access to these, simply click on the small triangle to the bottom left of the pane. Three separation amount figures can be adjusted to alter the separation of the panner. In the diagram above, the orange square represents the bounds of the possible separation of the panners. However, in some examples, it might be necessary to reduce this to reproduce a less wide image of the original. The three sliders correspond to the front left and right image (Separation XF), rear left and right image (Separation XR), and separation between front and rear (Separation Y). The fourth control sets the Level Compensation between "Unity Gain at Speaker," "Unity Gain at Mid Point," and "Legacy - Pre Logic 9.1.4.""

The LFE level, sometimes also called the low-frequency effect, is the "0.1" of 5.1. This is what is commonly known in the industry as a subwoofer. The LFE level control is placed within each panner as it is often undesirable to send everything to the subwoofer, when the main monitors should handle most bass information perfectly well. This again has come from the film industry where the LFE is used to reproduce rumbles, explosions, and impacts. The 5.1 standard means that all the main five monitors should be full-frequency and should reproduce bass well for most applications. Hence, the LFE is included to add weight to the bass end. However, it is worth remembering when mixing that for many audio applications, including home cinema systems, the subwoofer is not included as an "effect" but as the only bass generating driver, leaving the satellites to consider only mid- to high-frequency ranges.

Figure 10.34 In this example, the separation has been reduced to limit the width of the image. Also, it is possible to spread circular lines around the circumference that indicate which speakers are receiving a signal from which channel.

KNOWLEDGEBASE

Surround Sound Plug-Ins

Working in surround sound requires that you think slightly differently when employing effects such as reverberations. You could employ a stereo plug-in or you could choose to use two in tandem for front and rear if need be. Logic has fortunately thought of this and produced an intuitive set of surround-ready plug-ins to get you instantly working.

These dedicated plug-ins include the renowned Space Designer whose 5.1 algorithms are superb. Other 5.1 dedicated plug-ins include Delay Designer, Chorus, Tremelo, Flanger, Microphaser, and Modulation Delay.

The plug-in behaves in exactly the same way as the stereo version, but the response is returned in 5.1 as per the original impulse. To place a surround plug-in in your mix can be simply achieved by

(Continued)

selecting it from the menu. Most 5.1 plug-ins can be applied to a mono track whose output will be converted to surround. However, these processes are best added to a 5.1 bus or output.

Other plug-ins do not require specific surround algorithms to operate with multichannel information. These will, therefore, work in "Multi Mono" mode for processing multichannel information. Plug-ins such as the compressor, which utilizes this mode, will show some banks using additional switches located above the main controls that are not found on its stereo or mono counterpart. The Configure switch allows you to configure which compressor bank is attached to which outputs. Therefore, it is possible to strap a compressor across the front speakers, another to the rears, and one dedicated to the LFE channel. Alternatively, you might wish to alter this so that the Center channel has a dedicated bank for use in film work where the dialog is of paramount importance.

Figure 10.35 Choosing a surround plug-in can improve your 5.1 mix. Note how a mono track (or in this example a stereo track) will be processed with a surround sound plug-in thus changing its output to 5.1.

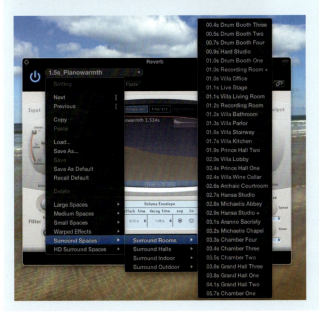

Figure 10.36 Space Designer comes with some impressive dedicated surround algorithms.

(Continued)

Figure 10.37 Plug-ins that do not require specific surround algorithms work in Multi Mono mode such as the compressor. Note the selector switches above the main plug-in to select which compressor bank operates which output.

Figure 10.38

10.7 Delivery Formats

Bouncing to Stereo and Surround

At some point, it will be necessary to present a mixed output of your production. If you are working within Logic, then it will be necessary to do a bounce, and depending on the application, different choices will need to be made. One of these choices will be of course whether to produce a stereo or a surround bounce. Bouncing surround mixes is achieved in the same way as traditional stereo mixes.

Once your surround sound mix is balanced and ready, simply select the "Bnce" button from the master channel strip in Logic's Mixer. A new dialog button will emerge showing a number of options for editing.

The first to note is the destination for the files once you have bounced them. Choosing the arrow to the top right will open up this dialog to show more of the file structure of the computer.

Figure 10.39
Bouncing to surround is as simple as clicking the "Bnce" button on the master fader strip. It is important to select the surround bounce checkbox and choose whether you wish the file to be interleaved or separated out.

Moving on to the left-hand side, the Destination table can be seen, which allows you to configure the file type you wish to export to. For most applications, this is likely to be PCM, which means unaltered or data compressed information, unlike the MP3 and M4A options below it. The DVD-A option is an excellent option for surround sound, as it will create a disc that can be played in most DVD players connected to a surround sound system.

Below this are the range controls allowing you to specify the start and end points of your bounce. This might be useful in the instance that you might need to export each scene separately for a dubbing mixer. Additional features include whether you wish to monitor the bounce in real time or allow the computer to manage the event offline. You are also given the opportunity to automatically normalize your audio as it is bounced.

The right-hand side of the dialog box includes a number of drop-down menus that relate to the file destination type you wish to export. First up is the option to choose the file type. There are three choices here starting with the Audio Interchange File Format (AIFF), which is generally the common format used on the Apple Mac. This file format is not subject to any data compression and is generally an accepted file format for use between software in the industry. AIFF does not contain timestamp information. Next is the Wave format or .wav and is perhaps the most common of audio file formats for exporting to another computer for further mixing or mastering. Finally, in the menu is the Core Audio File (CAF), which is an all-encompassing format that contains a wide variety of different formats in one and is common for Apple Loops.

The next drop-down menu is labeled Resolution and relates to the bit depth you wish for your bounce. This can be set typically at 8, 16, or 24 bits. Following this is the sample rate selector where a wide variety can be chosen from for a wide variety of applications, which will be determined by your destination. The File Type box allows you to set whether you wish for the files to be interleaved together or split files for each channel.

Dithering is required in situations where a lower resolution bounce is required than the project's set resolution – for example, when a 24-bit recording needs to be dithered down to produce a 16-bit CD ready file. Dither is a method by which the waveform is more accurately encoded for 16-bit reproduction and has been discussed previously.

The next, and perhaps most important feature for delivery of a multichannel export, is the Surround Bounce checkbox. This needs to be checked to enable that each channel of the surround field is exported as a separate file. The extensions for which are changed in the Audio Preferences as described before.

Exporting to DVD

Logic now allows you to export a DVD-Audio ready disc directly from the bounce dialog. Within this window, select "Burn to CD/DVD" and the right-hand pane should alter to show you the Mode. There are two choices here. DVD-A which can produce a 5.1 mix for use in DVD-A players or a (HD) stereo bounce of 24 bits and up to a 192-kHz sample rate. The CDDA option refers to our standard Red Book CD and as such higher sample rates and bit depths will be dithered to 44.1kHz and 16 bit.

For working to DVD Video, surround sound files need to be compressed using Dolby Digital Professional (AC-3) format. To achieve this, we need to use Apple's Compressor software which is part of Logic Studio or similar encoder software – see Knowledgebase.

Figure 10.40 Burning a DVD-A or CCDA can be achieved straight from Logic's own Bounce window.

First, we need to bounce the Surround Sound mix as split files, meaning that each stream (left, center, right, left surround, right surround, and LFE) is an audio file in its own right. This will then allow us to indicate in Compressor what files are for which output. Before bouncing, it might be worth checking the bounce extensions found in the audio preferences pane as discussed earlier.

Exporting Video with Audio (Dubbing)

Either during or at the end of the writing process for the film and television visual, you will be asked to send a copy with both the video and audio together into one file. This process is known in the industry as dubbing and is something that is normally handled by large postproduction houses, but for demo purposes, you can merge the two files together by selecting Export Audio to Movie from the movies menu (File > Movie > Export Audio to Movie …).

A new dialog box entitled Location and Name of New Movie will appear that allows you to choose the name and location for the file and the nature of the audio file. It is here that you decide upon the data compression format, sample rate, bit rate, and whether it should be stereo or mono. These settings are useful as a director or production company may wish to see your work in progress, and the best vehicle to deliver this to them is over the Internet via e-mail (if small) or File Transfer Protocol. It is, therefore, handy that the compression formats allow for a wide variety of standards to suit each and every client.

Once these settings are decided upon, click OK and choose the location for your final files. It is worth creating a folder at this point to ensure quick and tidy file management as there may be more than one file given the compression format you have chosen. Next click Save and wait while Logic compiles your video for you. Once completed, this file should be readable by Quicktime.

WALKTHROUGH
Creating AC3 Files Using Apple Compressor

For your new 5.1 mix to be read on a DVD-V disc, it will need to be compressed to Dolby's AC3 format. To achieve this using Logic Studio, you will need to export the six audio files (front left, center, front right, rear left, rear right, and LFE) using the "bnce" bounce feature. With six discreet audio files, it is now possible to compress for AC3.

To do this, make sure you've got your six audio files to hand. Quit Logic and search your applications folder for Compressor, purchasable from the App Store for £35 at the time of writing. Open this. You'll note Compressor looks quite different, but is a fantastic batch processing tool for all kinds of conversions and so on.

In Compressor, go to Job > New Job with Surround Sound Group.

A new dialog box will appear giving you options to place individual, discreet audio files in each speaker. You can add 5.1 or indeed an additional rear center channel "S" making this 6.1.

Also note the ability to add a movie here to compress the whole thing together for a DVD-V ready file.

Once you've clicked OK, you'll be back on the main Compressor screen and it is at this point you need to specify what the output file format will be. To do this, go to Window > Settings or simply click Cmd+3.

A new Setting dialog box will open that has two tabs: one for the Destinations of compressed file and one for the Settings. We're interested in the Settings at this point. There are a large number of file formats here to choose from, but seek out Formats > Audio > Dolby Digital Professional 5.1. Drag the Dolby Digital Professional 5.1 icon to the batch job in the main screen. Next press Submit on the bottom right to press your AC3 file.

Optimizing Logic

11.1   Introduction

Working with any music software will expect you to operate within its structure and design. Logic is part of that pack and offers a number of options to personalize the way in which you work with the application. One of the most exciting things about Logic is its flexibility and configurability to the way in which you need and like to work. Music production requires a fluid workflow that adapts to your way of making music. Doing this requires some initial thought and preparation, but it will improve not only the results but also the speed at which you work.

Over the years, Logic has developed an arsenal of features, tools, and techniques to help the workflow of productions by making the best use of the computer's interface. In this chapter, we'll explore some of those features and how to tweak Logic to fit your workflow.

11.2   Templates

As you launch Logic, you cannot fail to notice the chooser that allows you to get started with one of the program's in-built templates. Many of these are very workable and are an excellent starting point. However, as you begin to work in a particular way, you'll want your settings just so, thus saving your time and allowing you to be more creative. It is, therefore, worth preparing a template of your own with all the connections, plug-ins, windows, and other features ready to go to work.

It is best to spend some time thinking about the way this template will look and function. Consider the way in which you use Logic before setting up the ultimate template. You may need to set up different templates for doing different projects in a wide variety of genres. To do this, load up the Empty Project template from the "Explore" collection in the Template Chooser. Alternatively, you may just wish to alter a template from the chooser, which suits the way you work with your personal samples already preloaded into EXS24. Add to this project your external MIDI devices that you use frequently and give them tracks, your preferred organization of Screensets (discussed later), and then choose to save as a template (File > Save as Template…).

For example, if you do a lot of drum recording, with a guide bass, guitar, and vocals, it might be prudent to set up a template for this kind of recording, and perhaps, another template for working to picture. The benefits here are that all the tracks will be prelabeled and, therefore, your audio files too when you hit record. This can be extended further to consider the mix, and if you find yourself using the same plug-ins and settings for your components of the drums, the template can have these preloaded as in Figure 11.2. It is perhaps sensible to put these into bypass mode (Alt + click) as this can save on DSP power. In this example, the drum outputs are all set to go to Bus 3, which in this case has been set up as a Drum Group or Stem.

Figure 11.1
Logic's template chooser can save time when starting a project using a template of your own making.

Once you have created and saved your templates, they will appear in a new collection folder within the Template Chooser called "My Templates." These should provide you with personalized and honed platforms upon which to move forward fast with your projects.

Figure 11.2
Creating a template for a recording such as this can save time in the long run. Each strip can be automatically set up for sends and some with plug-ins as required. Ready to press "R" and record!

11.3   Screensets and Windows

Until Logic Pro 8, there were many individual windows to contend with. Each editor, mixer, list, and the Arrange Window were all separate and individually controllable windows. It's hard to believe now with the redesigned (first in Version 9) Main Window now updated for Logic Pro X with its "access all areas" philosophy. Logic's solution to managing all those errant windows remains and is still very relevant in certain circumstances.

Working in Logic may require that a significant number of windows be opened on top of each other at a time. The most obvious examples would be an Arrange Window and an editor opened at the same time, or the Arrange Window and Mixer together. Clearly, the current Logic Pro Main Window can deal with most of these adequately. However, if you wish to open other windows, most applications would use the Cmd + " ` " (the one to the left of the "Z" key) shortcut to allow you to toggle through the open windows in the program one by one. However, given the need to access so many different combinations of the Logic Pro Main Window, editors, mixers, and so on, Logic adopts what it called Screensets to allow for quick and easy navigation.

Screensets are Logic's ability to change what windows are viewed on the screen, or screens, at any time (a little like Mac OSX Spaces). On the title bar of Logic to the right-hand side of the Screenset menu, you will notice a number. This indicates the Screenset you have chosen, and pressing one of the numbers on the numeric keypad can change this, or you can click on the Screenset menu for more options. Within this menu, you have the opportunity to rename your Screensets, duplicate them, and delete them. The key additional features available within this menu are "Lock" and "Revert to Saved." These are really useful when you're using multiple monitors.

Lock allows you to set the way in which the screen looks so that it cannot be altered. A small bullet dot will appear to the left of the Screenset number to indicate this. This can be really useful when you need to rely on your Screensets being just so! There will be times when the Screenset changes you have made don't work for you and you wish to revert to your original decisions without losing the audio recording and editing you have done. To do this, choose the Revert to Saved option from the Screensets menu.

The window combinations can be carefully thought out to maximize both your speed and screen space. For example, most users will keep Screenset 1 as the default Logic Pro Main Window. However, whatever combination of windows you choose for each Screenset is entirely up to you. One example some users use is to follow the key commands for each window as a guide; see Window from the main Logic menu. Doing this means that you can get to navigate to the Screenset, which focuses on the task you need. For example, Screenset 8 (in Logic and before), could be set up to be the MIDI Environment, or the Audio Mixer within the Environment window, which we'll cover later in this chapter.

Figure 11.3

The Screenset menu not only gives you options to edit and work with Screensets but also offers a clear indicator from the menu bar of which Screenset you're currently working in.

Screensets really come into their own with two or more monitors. As we've already mentioned, the Logic Pro Main Window is a very comprehensive space for achieving most things in Logic, although there will be particular times when you'll want to see additional things, perhaps in larger windows or at the same time as something else. As such, Screensets could allow you, if you so desired, to manage the second monitor's content, while leaving the Logic Pro Main Window on the main monitor. Additionally, Screensets can also permit you to choose which editor or list is shown within the Logic Pro Main Window. For example, one Screenset could show the Sample Editor, while another shows the Markers list.

Should you wish to see the Mixer and the Logic Pro Main Window as in Figure 11.5, then this is absolutely fine. Although there is another more integrated way of bringing up the Mixer using the Environment window, Logic allows for the Environment window to be loaded up as a frameless floating window by selecting View > Frameless Floating Window in your Environment window. Before you do this, open a normal Environment window by going to Window > Environment, and change its view to "Mixer" using the arrow pointing down in the top left-hand side of the screen.

With this new floating Mixer window, shown below, it is possible to quickly navigate each track on the Logic Pro Main Window and the Mixer will follow as though the link control was on, making it an extremely flexible and powerful way of working efficiently whether on a small, wide, or double-screen setup. To enable this, you need to ensure that the link icon is enabled in the main Environment window before making it float. This might seem as duplication to the Inspector's expanded feature showing two Mixer elements, although you can, perhaps, be more dependent

Figure 11.4 The Windows menu offers all the other views that can be spread across additional screens should you not wish to use the integral panes in the Main Window.

Figure 11.5 In Logic X to create a floating Environment page, you need to select View > Frameless Floating Window.

on the number of monitors you have. The alternative use for this is perhaps where you might disable the link and focus the floating Environment window (as a mixer) to the Master output. Now, directly from the Logic Pro Main Window, we operate the focused track's Mixer element in addition to the master bus.

Immediately, there are nine Screenset levels that can be accessed from the number keys on both the qwerty and numeric keypads. However, Logic allows you to access and make use of Screensets up to 99.

Figure 11.6 Two windows can be made to sit side-by-side using Screensets as shown here, but it is also possible to create a floating Environment window (a mixer layer in this instance), which can be linked to your actions on the Tracks Area (below).

To key in these Screensets, you need to press Ctrl and type the number of the corresponding Screenset. So for 34, simply press Ctrl + 3 and then Ctrl + 4, on the numbers on the qwerty keyboard.

At times you may receive a file you have been asked to mix, which will have different Screensets and perhaps different key command settings. It is important that your workflow is not jeopardized and that you can work both efficiently and productively with your preferences. To import project settings from another Logic File, go to File > Project Settings > Import Project Settings.

Figure 11.7
It is possible to import your project working preferences from another Logic File using the Import Settings feature.

Link Modes

Link mode when mixing can be really useful so as not to overload one screen with multiple plug-ins.

Link has two discrete modes. The first is Same Link Level (pink chain link icon), which means that whatever item you click on, whether that be a region or the note, the linked windows will follow the same track. These differences are ever so subtle, and frankly, it is one of those things that if the windows are not linking as you need them to, then simply press Ctrl + click the link icon, change the mode, and see if that works the way you want it to. Nevertheless, the Link mode is a key feature of Logic as it does ensure that Screensets can operate fluidly.

11.4 Key Commands

Keyboard shortcuts are a useful way of improving the workflow with whatever software you are using from a word processor to an Internet browser. Logic, of course, is no different, and as you'd expect, there are so many of them covering a vast number of the features. Throughout this book, we give you the new default key commands that relate to what we have covered in Logic Pro X. Most of these commands can be learnt by noting the writing to the right of the command in one of the menus, but the whole range can be viewed in their own dedicated Key Commands viewer (Logic Pro X > Preferences > Key Commands… or Alt + K).

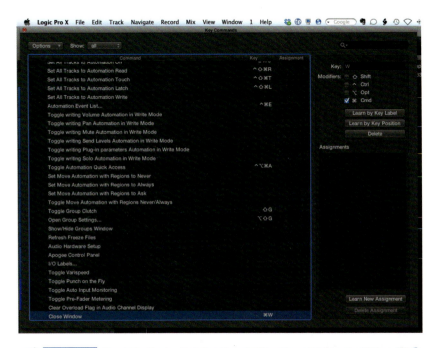

Figure 11.8
(a) Key Commands are an important feature of any piece of software to increase productivity. Logic is no exception allowing you to carve out a key command for literally every feature. (b) For those not yet brave enough to learn the new shortcuts as well as the new software will be pleased to engage the Logic 9 keyboard shortcuts.

Searching for Key Commands is easy by using the Spotlight-like search bar on the top right of the pane. Here, simply type in the command you wish to find the key command for and the possible results will appear under one of the main headings in the command list. Under each of these 17 headings, are all the key commands available for use with Logic. Not all of these commands will have a shortcut associated with them as there are simply too many. But as with all things in Logic, they too can be optimized or altered to suit your working needs.

Once you have selected the instruction you wish to create a key command for, select it and choose to either "Learn by Key Label" or "Learn by Key Position" followed by the key combination you wish to use. There are subtle differences between these buttons. Logic allows you to discern between using the numerical keypad and using the numbers on the main part of the keyboard. In this way, you could choose to assign two different commands to number 9; one would be on the main keyboard and the other on the numerical keypad. Therefore, learning by Key Label means that using the number 1 will call up the same key command whether it is on the numerical keypad or above the qwerty. However, the Key Position option will allow you to discern between the two differing positions, thus giving you further key command possibilities.

There will, perhaps, come a time when you will develop your own key commands and see your productivity suit the way in which you work. Your templates will already preload your key commands with your work if you saved them, but there may come a time when you have to work on an imported project such as a mix that someone else put together. In this instance, it is really important that you save a set of your personalized key commands.

There will be times when you may need to impose your special and personalized changes to another's Logic session and will need to have your key commands ready. To do this, we need to export our key commands. To do this, simply press Alt + K from anywhere in Logic to bring up the Key Commands dialog. Within the Options menu is a command called Export Key Commands that will allow you to save your key commands. These can then easily be exported into another copy of Logic using the Import Key Commands instruction in the same menu. These key commands can also be saved to the clipboard for use as a reference, or for revision. It may even be worth doing this while you're learning to use these commands.

If, however, you have learnt Logic's default Key Command set and import another person's project, you might be at a loss as to why so many of them did not work. In this case, it would be sensible to choose the option of "Initialize All Key Commands," which will reset the commands to the factory standard.

11.5 The Environment

Logic's Environment window has an interesting reputation to the non-Logic using world. It has always held an almost mythical-like status, where many users have not wished to delve for the fear of it would mess their whole project up. Although the Environment is perhaps less used these days now that MIDI is not the main method of communication between the application and the sound sources, it is still an important tool that allows the user to experiment and perhaps optimize his or her creative output. The Environment window is thus the route for information into and out of Logic and as such is something that is worth considering especially when working with MIDI. It is here that the virtual representation of your studio can be created, managed, and manipulated. In this section, we'll give a quick overview of its core functions and its possibilities. Where you take it will be up to your creativity.

Layers and Windows

The Environment window itself could be simply looked upon as an area to organize your MIDI inputs and outputs and how they all connect together. However, there are some more angles, or layers, to it which we'll explore a little here, such as its Audio Mixer and Global Object layers. In essence, the Environment should show all the connections to and from Logic, both physical and internal.

The first layer in the list is the All Objects list, which shows you all the connections to Logic currently established. This includes MIDI, Audio, and internal connections such as ReWire. The next layer is Global Objects that allows you to specify any objects that are common to each layer. The Click and Ports layer refers to how the click track (metronome) is managed within the MIDI environment, and how the Ports connect to each MIDI input, whether that be physical through an interface (port) or the Musical Typing Keyboard. The MIDI Instr. or MIDI Instruments layer is to allow you to specify what each instrument is and to manage its operation from within Logic. The last is the Mixer layer, which allows you to create a customizable mixer layout to suit your way of working.

Figure 11.9 The Environment has a number of different layers, which can be selected using the menu at the top left-hand side of the window.

Figure 11.10
The Environment has a number of different layers, which can be selected using the menu at the top left-hand side of the window.

New layers can be created for different purposes by selecting "Create Layer" from the menu. It might be prudent to create perhaps an additional layer if you wished to use the Environment to control an external MIDI synthesizer using something we call a mixermap, or a controller. This map could allow a physical display of the controls of that MIDI synth within a new Environment layer. The benefit here is that any movements you make to the synth within Logic can be recalled and automated. What you

Figure 11.11
The Inspector is an important aspect of the Environment window and can be toggled in and out of view using the "I" key.

choose to use on additional layers will depend on the way you work. Additional mixer layers might be an idea, one for the channels and another for auxiliary tracks, for example.

An additional dimension to the "layout" of your windows is the opportunity to make the Environment window a Frameless Floating window by selecting it from the View menu. If you press link on the Environment window before making it frameless, it will follow the regions you select as you move around the project. As the Environment window can be used for so many specialized and personalized facilities, it might be necessary to have it open on top of other windows. In this instance, create a Frameless window (View > Frameless Floating Window) and place it in your Screenset next to your Logic Pro Main Window or however you see fit for your session management.

Objects

Each element within a layer is known as an object, These objects and their connections make up the manner in which Logic works. To see the types of object available, go to the New menu within the Environment window where you will be greeted by a long list. The list is grouped into four. The first is the objects themselves; the second is a collection of processors, such as the Arpeggiator or Transformer. The third collection comprises simply two inputs: Physical Input and Sequencer Input. The Physical Input refers to the MIDI input from an external device, perhaps a master keyboard, whereas the Sequencer Input refers to what Logic "gets." In other words, one could place a process such as the Arpeggiator between the Physical Input and the Sequencer Input. The last collection refers to Audio and external functionality such as new Internal connections, which include ReWire, and Audio objects such as auxiliary tracks.

Managing objects is simple as they can be moved, copied, and edited in a way similar to that of regions on the Tracks Area. The Environment window also has an Inspector just as the Logic Pro Main Window and can be toggled in and out of view using the "I" key. Selecting an object will reveal some editable features within the Inspector.

It is quite easy to create some unique layouts such as mixers, which would benefit from being presented in a structured and uniform manner. When creating an environment, it is easy for the objects to not quite line up. As with many graphics applications, Logic's Environment also includes a snap to grid-type facility called Snap Positions (in the Environment, View > Snap Positions). This will ensure that the objects you place can be easily managed and look smart!

Cables

Objects need to be connected to operate in many instances. Therefore, Logic provides "cables" to connect one object to another and can, therefore, be extremely flexible.

If you navigate to the Click & Ports layer, you'll notice that the Physical Inputs box has a number of white triangular arrows running down the right-hand side. These are connected to some writing that indicates which physical MIDI input they refer, such as Port 1, and so on. Alternatively, at the top of this box is a SUM of all the physical inputs (which can be really useful if your master keyboard is away from your computer as you could employ the Musical Typing Keyboard or another smaller keyboard) and either signal will be passed into Logic.

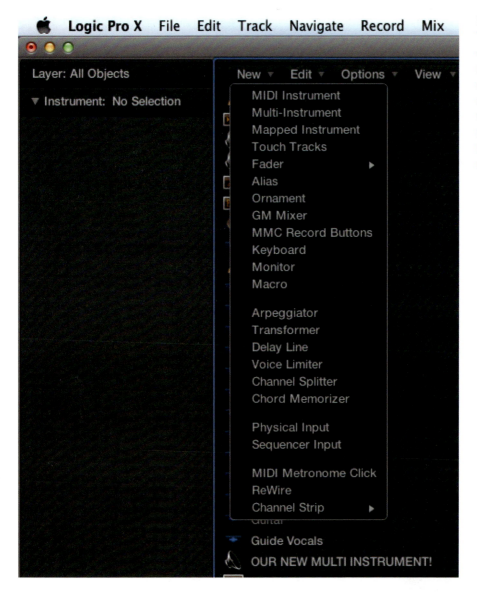

Figure 11.12
There are a considerable number of environment objects available to you to choose from to control Logic in a flexible manner.

It is likely that the SUM output of the Physical Input box has, by default, been connected to an Input Notes keyboard on screen, which allows you to click and preview notes as though you were playing a keyboard. This is usually automatically connected to what is labeled as the Input View, but is actually known as a Monitor Object. This shows you the MIDI information as you play the keyboard, which is usually a Note On message followed by a Note Off for each key pressed. MIDI information then travels through the cable connected to the Sequencer Input, which connects to your chosen track on the Logic Pro Main Window.

As a default, this is great as it allows you to see how the Environment connects, and it is at this point that you perhaps might wish to experiment a little. For example, try creating an Arpeggiator and another Monitor Object. The idea here is to place the Arpeggiator after the original Input View Monitor to create a new musical arrangement. This could then be outputted to another Monitor Object to see what changes have been made to your input. With this arrangement of objects, anything played in on any physical inputs will be subjected to an Arpeggiator.

The Arpeggiator will run at the tempo of the project, and its response and range can be altered by selecting it in the Environment and visiting the Inspector to the left of the screen.

As Click & Ports generally refers to the input side of Logic, there is so much experimentation we can do here. There are obviously many more options: we could try working on the output of Logic, which can be quickly demonstrated by moving the mixer layer and adding in a new Arpeggiator into the environment. Connect this to a software instrument channel at the top of the fader. This new Arpeggiator object on the Mixer will now be available on the Logic Pro Main Window. Create a new track and press Ctrl + click the track header to open up the Reassign Track Object menu. From here, select the Mixer submenu and you should see the word Arpeggiator written in. This will now allow you to record and play only the source notes, which will then be arpeggiated each time the sequence is played.

Figure 11.13
Using the Environment, it is easy to create unique and interesting creative aspects to your work. In the examples above, we have placed an Arpeggiator across any MIDI input to Logic.

Figure 11.14
The Environment has a number of different layers, which can be selected using the menu at the top left-hand side of the window.

It is quite possible that the cables on an environment may get so frenetic that you'll not be in a position to see the objects! Within each layer is the option to view or hide the cables, thus allowing you to see what is going on. Simply go to the View menu and select Cables. This will hide them until the next time you need them. To make the workflow quicker, you can simply press Ctrl + C to toggle them in and out of view.

Another dimension is the ability to color your environment cables by clicking on the source, that is the device that is outputting to the cable, and pressing Alt + C or View > "Colors…." Change the color as necessary to enable quick access to the appropriate element of your environment. Remember these can simply get toggled out of view if it all becomes too much.

Setting Up MIDI Instruments in the Environment

The MIDI Instruments Layer is an important one as it allows you to specify the MIDI environment and the synths and devices you have attached to your system. For example, if you bring in a new MIDI synth to your setup, you could simply just tell Logic there is a new instrument connected, and this will allow you to select it from the track assignments, which is fine, but there is so much more that can be done.

Creating a new instrument within the MIDI Instruments layer will enable a new instrument to be located within the Reassign Track Object menu meaning that you can directly select the device and its channel from the Logic Pro Main Window. To do this, we need to create a new Environment Object, and there are two main alternatives for MIDI instruments. The first is to create a new "Instrument" (New > Instrument). This means that this output will go to one MIDI channel from any port. There are some classic examples of MIDI instruments that are not multitimbral such as the Waldorf Pulse, which is a single MIDI-channeled monosynth.

For larger multitimbral synthesizers, we'll need to create a "Multi-Instrument." This can allow us to specify quite a lot about the device in question. As you select this, you're confronted by a box with 16 small buttons. These will all be crossed off. This is essentially to save your time as not all multitimbral

Figure 11.15

It is easy to view, hide, or protect the Environment's cables by going to View > Protect Cabling/Positions.

synthesizers are able to use all 16 MIDI channels. Therefore, you can specify which outputs you wish to use. Another reason for limiting the output of certain channels might be because you may wish to use a channel from the same physical output, if your MIDI outputs are limited, to send to a monosynth such as the pulse on channel 16.

After creating an instrument object, you will need to connect it to an output port. To do this, simply select the instrument in question and visit the Inspector to the left of the Environment window. Usually, directly under the Icon check-box is a drop-down menu that lists all the available physical ports. Simply select the physical MIDI port you have connected the device to, and you can use the instrument straight away from the Logic Pro Main Window. This is known in Logic speak as a direct output assignment.

Once you have set the device using the multi-instrument object, you can double-click the object to reveal the Multi-Instrument Window. This immediately shows the instrument names for a General MIDI device (GM Device) that relate to a program change that can be sent by the sequencer. The idea is that you can relabel the instruments within Logic as they relate to their corresponding program change number. Once this has been done, you can choose the instrument name when selecting the Track Assignment in the Logic Pro Main Window. Although this might seem like a long-winded exercise for all your esoteric synths, the operation would only need to be completed once. That level of recall from the Logic Pro Main Window is an excellent feature and once done can speed up your workflow without leaving your mix position.

The Mapped Instrument Object can be useful when you wish to convert one type of MIDI input data to another. For example, you may have drum kit loaded in an external device, which does not follow the General MIDI drum convention. In that instance, it would be useful to call up a Mapped Instrument Object that can be double-clicked to open a chart, which allows one note to be mapped to another in real time.

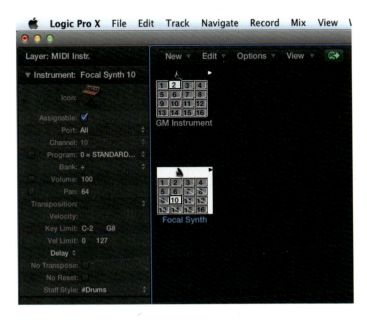

Figure 11.16

Creating a new Multi-Instrument in the MIDI Instruments layer will allow you to create a new stream of MIDI information intended for this new device.

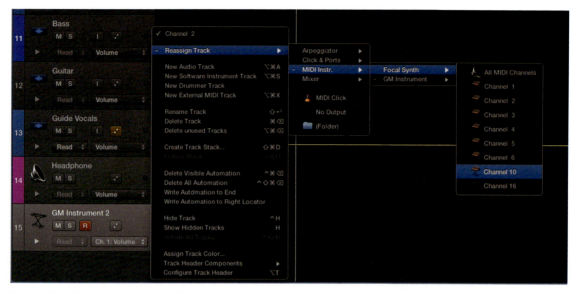

Figure 11.17 Creating a new Multi-Instrument in the MIDI Instruments layer will allow you to create a new stream of MIDI information intended for this new device.

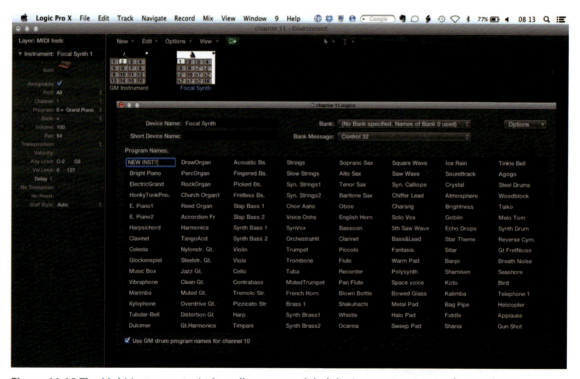

Figure 11.18 The Multi-Instrument window allows you to label the instruments or patches to their corresponding program change. This can be then selected from the Program Change menu in the Inspector.

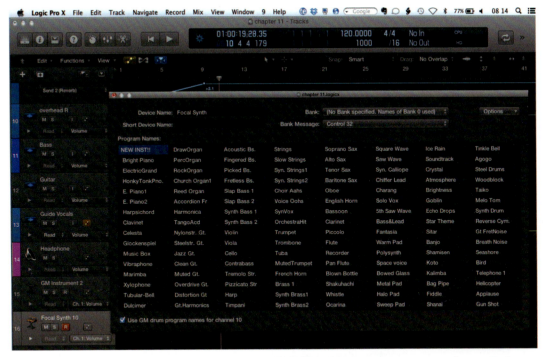

Figure 11.19 The Multi-Instrument window allows you to label the instruments or patches to their corresponding program change. This can be then selected from the Program Change menu in the Inspector.

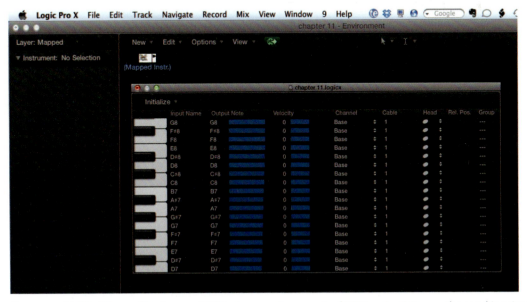

Figure 11.20 The Mapped Instrument Object converts one set of MIDI notes into another and is primarily useful for mapping drum kit sounds.

WALKTHROUGH
Adding New MIDI Devices to the Environment
Step 1

Adding a new MIDI instrument to the Environment will allow you to select it as a distinct item from the Reassign Track Objects menu. First, open up the Environment window, which is Cmd + 0 or Windows > Environment. With the Environment open, select the layer selection menu to the top left. From this menu, select the MIDI Instruments layer abbreviated to "MIDI Instr." This layer should see a GM Device, which is sometimes routed to the Quicktime Music Player.

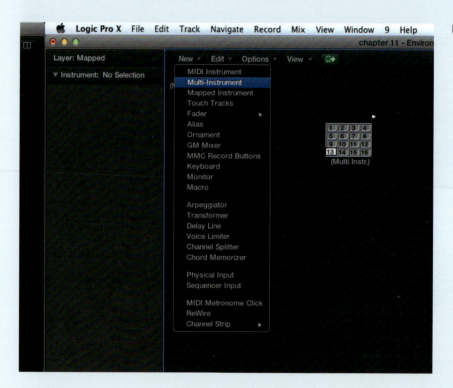

Figure 11.21

Step 2

To create a new instrument, choose the New menu. There are plenty of options available here, but namely we're concerned with either a new Instrument or Multi-Instrument. An "Instrument" in this case is an instrument that has only one MIDI channel, such as synthesizer. The Multi-Instrument alternative is for use with multitimbral MIDI devices such as a sound module or external sampler. It is important to establish which type your device is before connecting as it will make your selection easier from a track in the Logic Pro Main Window. When this appears, you will need to route it out of a physical MIDI output. To do this, select the instrument object and visit the Inspector on the left to select the port it will output from.

(Continued)

Figure 11.22

Step 3

If you have chosen the Multi-Instrument option, a grid with 16 numbers will greet you, and they'll all be crossed out. This is so that you can specify how multitimbral your device is. For example, some synths only can receive eight channels of MIDI, rather than the full quota of 16. In this instance, simply don't deselect the last eight. This will ensure that Logic can only talk to the eight live channels. While you're here, it is really handy if you label the instrument by either selecting the object and clicking on its name in the inspector or selecting the Text Tool from the Click Tools menu. Your device will be labeled and available from the track assignments in the Logic Pro Main Window.

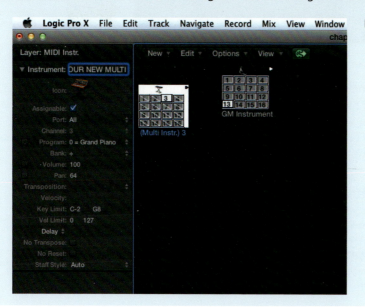

Figure 11.23

Live MIDI with the Environment

Earlier, we spoke of the Arpeggiator in terms of cabling and the creative effects it can have. The sheer number of objects possible within the Environment window allows us to design some interesting and important facilities for working with MIDI. Whether that is an Arpeggiator or Multi-Instrument, the two together can make some interesting outputs. This can, of course, be extended further if the fancy takes you. In the couple of examples below, we look at some aspects that can make Logic a useful tool when working live.

There are some interesting elements to the software, which might be worth mentioning here. First is the Chord Memorizer Object whose job in life is to do a more clever version of what many lower end keyboards are renowned for: one-note chords. The concept is that you can detail what the input note will be and what chord results from it. To create this, simply visit the View menu and click on a Chord Memorizer Object. Next, double-click the object itself to reveal two keyboards that allow you to set which chord plays depending on which key would be pressed.

Figure 11.24
Chords can be memorized to be played using one key using the Chord Memorizer.

The Touch Tracks Object offers the Logic user a range of triggers to be enabled to play certain regions when pressed, like one of the functions of Ableton Live. This is a process that is limited to MIDI regions and Folder Track Regions. Select a Touch Tracks Object from the New menu and double-click it as it comes into view. Another table appears to allow you to assign a region or folder region to certain keys. This can be incredibly flexible when working live.

External Control with the Environment

One of the most flexible aspects of Logic is its ability to control external equipment which responds to MIDI. Whether to an external analog or a digital mixer that responds to MIDI messages or a synthesizer, Logic can control whatever functionality it allows using something termed a "Mixermap." This can be extremely flexible to allow you to further control your equipment from the mix position. Any moves you make can be saved with the project as the MIDI can be saved as automation.

Input Name	Group	Region/Folder	Transpose	Velocity	Trigger	Start	Delay
G8	Off	(unassigned)	⇕	100% ⇕	Gate	⇕ Free	⇕
F♯8							
F8							
E8							
D♯8							
D8							
C♯8							
C8							
B7							
A♯7							
A7							
G♯7							
G7							
F♯7							
F7							
E7							
D♯7							
D7							
C♯7							
C7							

Figure 11.25 Touch Tracks enables Logic users to work with a range of MIDI regions mapped to various keys. Great for a live performance!

To enable this, simply create a new layer, which can be renamed as the device you're planning on automating. Within this layer, create a series of objects which reflect the device and the controls you want to manipulate. Each object will have a different function to manage in the destination device. To alter this, we need to select each object and alter the MIDI message type it sends out. In the example on p. 392, we wish this fader to manage a volume; hence, this should be control change 7 on MIDI channel 1. This would need to be repeated for each channel and message for the Mixer.

There will, however, be instances where the MIDI data required will need to be somewhat more complex, such as System Exclusive (Sys EX) which will be required for the mixermap. For these message types, it will be trickier to enable as this takes more knowledge of MIDI than the remit of this book, but is of course possible. If your external synthesizer follows General MIDI, then simply call up Logic's own GM Mixer by creating a new layer and then visiting New > GM Mixer. This might be a good starting point for the creation of any new mixermaps.

When creating mixermap environments to control externally connected MIDI instruments, it is worth noting that these are best designed on the output side of the Environment. By "output side" we mean by cabling up the faders directly before introducing a physical output using an instrument object as can be seen in Figure 11.26.

There is a bit of a strange logic to this, but doing it this way, you immediately take control the device from the mixermap by connecting the fader objects through to the MIDI instrument. Logic immediately records any actions you make on the environment page should you wish to without the need to cable these object to another sequencer input.

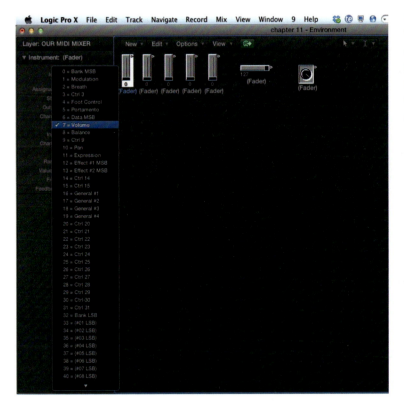

Figure 11.26
To change the assignment of the fader, simply go to the Inspector and select the type of message and the corresponding number. In this case, we should send out a volume message.

Your MIDI device, which might be a mixer as in the environment example below, may also be able to transmit MIDI commands as you alter its physical controls. If so, these can be, in turn, fed-back into Logic and be reflected on the Environment's mixermap, provided it is cabled correctly. This loop is necessary to ensure that both the physical and the virtual controls are representative of each other. Fader objects on the Environment will only be controlled, though, if the physical device is routed back to the mixermap.

To do all this, simply create a new layer and name it "Mixermap" or something similar. Create an instrument (New > MIDI Instrument), which will act as your "Physical Output" directly to the MIDI-controlled device, in this example a mixer. It is worth ensuring that the icon checkbox is unchecked for this object as this does not prevent the icon from showing in the environment (we've changed it to a MIDI Plug here), but it will no longer show up in the Tracks Area when selecting outputs.

Next select the faders and other object types that best describe how you wish to control the external device. Select each new object and specify its parameters, such as control change number 7 for volume and control change 10 for pan, if it were a GM Device. It is worth noting at this point that many MIDI devices use different message types than you'd expect, so it's best to visit the corresponding manufacturer's manual. Alternatively, put Logic into record and move the appropriate fader on the MIDI device. Once recorded, simply visit the Events List (Window > Events List or Cmd + 0) to ascertain what control change message it transmits. Later, we'll show another easier way to do this in the Environment window.

Figure 11.27
When making
mixermaps, it is
advisable to cable
your fader objects
directly to a physical
MIDI output using
an instrument
object.

In order for this to operate effectively, you need to cable up the faders and other objects to the newly created instrument we've called "Physical Output." To the right-hand side of each object are some triangles that represent cable "sockets." Simply click here and drag the mouse away to show some virtual cables, which can be dropped onto other objects to make a connection as in the example above.

These cables then need to be connected to the newly created "Physical Output" to transmit the changes you make on screen to the device. This should work provided the MIDI interface and MIDI device are operating properly.

However, we need to ensure that any movement on the physical controls on the MIDI device is reflected on your Environment Mixermap. To do this, we need to make a connection between the physical input to the Environment and the Mixermap layer. First, it is necessary to add a new monitor object to the newly created Mixer. This Monitor object will act as both a way of visualizing the MIDI information as we would in the Events List and as a method of connecting layers together. Connect the Monitor's output to the first mixer object.

We then need to create a new instrument object on the Clicks & Ports environment page if you have not done so already for this MIDI device. In the example above, we've named it "Our MIDI Mixer." If this were a synthesizer or sound module, then this will appear in the Tracks Area track lists as usual.

To get the Mixermap Environment layer to respond, we need to cable this instrument object on the Clicks & Ports layer to the mixermap itself. Simply select a cable output from the instrument object on the Clicks & Ports layer, and with both layers open simply drag and drop the cables between layers to the Monitor Object. An easier way would be to press Alt + click the output of the instrument object which reveals a menu similar to that found on the track list on the Tracks Area. This second method will only work, provided you have enabled the "Icon" check box which relates to the mixermap's Monitor Object.

Provided all the objects are cabled correctly, your newly created mixermap should offer real flexibility to control devices directly from your Mix position. The open-ended nature of the Environment allows for many more possibilities to control devices within your studio.

11.6   Input/Output Labels

When working within a large studio environment that brings together a fairly big analog setup with many processors and a mixing console, there may be simply too many connections for you to remember which specific input relates to your prized Fairchild unit. Like with so many things within Logic, this can of course be labeled upon some physical inputs. To do this, select Mix > I/O Labels ... A new window opens showing all the physical inputs and auxiliaries within the system.

Figure 11.28 The Input/Output (I/O) Labels within Logic can be named to improve the workflow of your session. Naming the key inputs and outputs and auxiliaries makes things quicker.

To change the names of connections, click either in the Long Name or in the Short Name boxes. The Long Name is your description and should be whittled down to a smaller amount of characters for the Short Name which will be used more frequently when accessing drop-down menus in the Mixer.

Figure 11.29
The changes made in the Input/Output (I/O) Label dialog box makes all the difference with respect to quickly recognizing busses in the Mixer.

Index